Microbial Risk Analysis of Foods

Emerging Issues in Food Safety
SERIES EDITOR, Michael P. Doyle

Microbiology of Fresh Produce
Edited by Karl R. Matthews

Microbial Source Tracking
Edited by Jorge W. Santo Domingo and Michael J. Sadowsky

Microbial Risk Analysis of Foods
Edited by Donald W. Schaffner

ALSO IN THIS SERIES

Enterobacter sakazakii (2007)
Edited by Jeffrey M. Farber and Stephen Forsythe

Foodborne Viruses: a Multidisciplinary Review of Progress and Challenges in Foodborne Virus Research (2008)
Edited by Marion Koopmans, Dean O. Cliver, and Albert Bosch

Safety of Imported Foods: Microbiological Issues and Challenges (2008)
Edited by Michael P. Doyle and M. C. Erickson

Microbial Risk Analysis of Foods

EDITED BY

Donald W. Schaffner
Department of Food Science
School of Environmental and Biological Sciences
Rutgers University
New Brunswick, New Jersey

ASM PRESS

WASHINGTON, D.C.

Address editorial correspondence to ASM Press, 1752 N St., N.W., Washington, DC 20036-2904, USA

Send orders to ASM Press, P.O. Box 605, Herndon, VA 20172, USA
Phone: 800-546-2416; 703-661-1593
Fax: 703-661-1501
E-mail: books@asmusa.org
Online: http://estore.asm.org

Copyright © 2008 ASM Press
American Society for Microbiology
1752 N St., N.W.
Washington, DC 20036-2904

Library of Congress Cataloging-in-Publication Data

Microbial risk analysis of foods / edited by Donald W. Schaffner.
   p. ; cm.—(Emerging issues in food safety)
   Includes bibliographical references and index.
   ISBN-13: 978-1-55581-461-8 (hardcover : alk. paper)
   ISBN-10: 1-55581-461-1 (hardcover : alk. paper)
   1. Food—Microbiology. 2. Food—Analysis. 3. Food—Risk assessment. I. Schaffner, Donald W. II. Series.
   [DNLM: 1. Food Microbiology. 2. Food Analysis. 3. Risk Assessment. QW 85 M62617 2008]

QR115.M457 2008
664.001′579—dc22

2007016819

10 9 8 7 6 5 4 3 2 1

All Rights Reserved
Printed in the United States of America

Cover illustration: *E. coli* microcolony. Courtesy of James A. Shapiro, University of Chicago. Licensed for use, ASM MicrobeLibrary (http://www.microbelibrary.org).

Contents

Contributors vii
Series Editor's Foreword ix
Preface xi

1 Qualitative Risk Assessment 1
Marion Wooldridge

2 Using Risk Assessment Principles in an Emerging Paradigm for Controlling the Microbial Safety of Foods 29
Richard C. Whiting and Robert L. Buchanan

3 Microbial Ecology in Food Safety Risk Assessment 51
Tom Ross

4 The Modular Process Risk Model (MPRM): a Structured Approach to Food Chain Exposure Assessment 99
Maarten J. Nauta

5 Using Risk Analysis for Microbial Food Safety Regulatory Decision Making 137
Sherri B. Dennis, Janell Kause,
Mary Losikoff, Daniel L. Engeljohn, and
Robert L. Buchanan

6 Integrating Concepts: a Case Study Using *Enterobacter sakazakii* in Infant Formula   177
Martine W. Reij and Marcel H. Zwietering

7 Communicating about Microbial Risks in Foods   205
William K. Hallman

Index   263

Contributors

Robert L. Buchanan
Center for Food Safety and Applied Nutrition, Food and Drug Administration, 5100 Paint Branch Parkway, College Park, MD 20740-3835

Sherri B. Dennis
Department of Health and Human Services, Food and Drug Administration, Center for Food Safety and Applied Nutrition, HFS-006, 5100 Paint Branch Parkway, College Park, MD 20740-3835

Daniel L. Engeljohn
United States Department of Agriculture, Food Safety and Inspection Service, 1400 Independence Avenue, S.W., Room 349-E, Washington, DC 20250-3700

William K. Hallman
Food Policy Institute, Rutgers, The State University of New Jersey, ASB III, 3 Rutgers Plaza, New Brunswick, NJ 08901-8520

Janell Kause
United States Department of Agriculture, Food Safety and Inspection Service, 1400 Independence Avenue, S.W., Room 333 Aerospace Center, Washington, DC 20250-3700

Mary Losikoff
Department of Health and Human Services, Food and Drug Administration, Center for Food Safety and Applied Nutrition, HFS-415, 5100 Paint Branch Parkway, College Park, MD 20740-3835

MAARTEN J. NAUTA
Microbiological Laboratory for Health Protection (MGB), National Institute for
Public Health and the Environment (RIVM), P.O. Box 1, 3720 BA Bilthoven,
The Netherlands

MARTINE W. REIJ
Laboratory of Food Microbiology, Wageningen University, P.O. Box 8129, 6700
EV Wageningen, The Netherlands

TOM ROSS
Australian Food Safety Centre of Excellence, School of Agricultural Science,
Private Bag 54, University of Tasmania, Hobart Tasmania 7001, Australia

RICHARD C. WHITING
Center for Food Safety and Applied Nutrition, Food and Drug Administration,
5100 Paint Branch Parkway, College Park, MD 20740-3835

MARION WOOLDRIDGE
Centre for Epidemiology and Risk Analysis, Veterinary Laboratories Agency
(Weybridge), New Haw, Addlestone, Surrey KT15 3NB, UK

MARCEL H. ZWIETERING
Laboratory of Food Microbiology, Wageningen University, P.O. Box 8129, 6700
EV Wageningen, The Netherlands

Series Editor's Foreword

Risk analysis is a tool to facilitate complex problem solving and decision making in addressing mitigation of a hazard and comprises three components. These include risk assessment, risk management, and risk communication. Risk assessment involves using scientific information to describe the likelihood and magnitude of harm attributed to a specific hazard. Risk management applies to the activities undertaken to control a hazard. Risk communication is the exchange of information about a hazard among interested parties. Several monographs have been written describing the principles of these three concepts; however, few have been published regarding how these concepts apply to today's initiatives to provide safer foods.

Microbial risk analysis has become an integral part of a science-based decision-making process to identify intervention strategies that yield the greatest impact in providing public health protection to the food supply with the limited resources available. A recent example is a risk assessment of *Listeria monocytogenes* in foods conducted by the Food and Drug Administration and the U.S. Department of Agriculture that has enabled the agencies to risk-rank foods having the greatest potential in serving as vehicles for listeriosis. These regulatory agencies have applied this information to risk management to focus their resources on foods of greatest risk, such as deli meats, and away from foods that are highly unlikely to cause illness if consumed, such as ice cream. Furthermore, this information, through risk communication, has been disseminated to food processors and consumers to enable them to apply greater interventions to high-risk foods. The net result is the prevention of more cases of listeriosis than would occur if all foods were equally targeted with the same amount of resources.

The concept of microbial risk analysis is not only used in the United States but has taken on international dimensions, with the World Health Organization and Food and Agriculture Organization of the United Nations actively engaged in conducting risk assessments and applying them to the risk analysis matrix. This information has subsequently been used by Codex Alimentarius in its deliberations for setting international food safety standards. Hence, microbial risk analysis has become a tool adopted worldwide to mitigate food contamination and related illnesses.

Written by a cast of internationally recognized authorities, this book elucidates how risk analysis of food-borne agents of microbial origin can be used to provide greater public health protection to our food supply. I commend Don Schaffner and his associates for this extraordinary contribution.

MICHAEL P. DOYLE, Series Editor
Emerging Issues in Food Safety

Preface

Microbial risk analysis is a new and emerging component of the field of microbial food safety. It is being used with increasing frequency by domestic food safety agencies as well as by international organizations to assess and manage the safety of our food supply.

There are several reasons why microbial risk analysis has emerged as such an important food safety tool at this point in time. First, the advances in computing technology have made powerful risk analysis tools accessible to a wider array of scientists and engineers than ever before. Second, our food production and distribution systems have become quite complex. So much so, in fact, that it may not be intuitively obvious exactly where the optimal food safety fixes are needed, unless a tool like risk analysis is used.

Although these two factors have made the development of microbial risk analysis possible, and even desirable, the most significant factor driving the international interest in risk analysis is one related to international policy and trade. A component of the World Trade Organization (WTO) Agreement on the Application of Sanitary and Phytosanitary Measures (called the SPS Agreement) deals with risk. Sanitary measures are those that relate to human and animal health (e.g., salmonellosis or foot-and-mouth disease) while phytosanitary measures relate to plant health (e.g., Mediterranean fruit fly). Article 5 of the SPS Agreement is entitled "Assessment of Risk and Determination of the Appropriate Level of Sanitary or Phytosanitary Protection." The first paragraph of Article 5 states that sanitary or phytosanitary measures should be based on an assessment of the risks and should take into account risk assessment techniques developed by the relevant international organizations. Because of the SPS Agreement, many importing and exporting countries are interested both in carrying out risk assessments of domestically produced foods and in understanding the risk posed by imported foods.

Because of the advances in computing technology, the complexities of our food production and distribution systems, and the need for fair international trade, we have seen a dramatic increase in the interest in microbial risk analysis as well as in its practice, both by national governments and international bodies like the World Health Organization and the Food and Agriculture Organization. It is within this context that I was asked to edit this important book.

While many texts are available that describe the basics of microbial risk assessment, a real need exists for an "advanced" text on the subject. The basic idea of this book is to answer the question: what do you need to know after you understand the basics of microbial risk assessment? This book has been written primarily for beginning risk assessors who need to take the "next step" in their education. It will also be a valuable reference for the graduate student or university researchers engaged in risk assessment research, or serve as a text for a graduate level course in microbial risk analysis. It will be a valuable resource for industry and government risk managers who need to understand the sort of information their risk assessors will be communicating to them in the next five years.

Topics covered in this book include qualitative risk assessment; the relationship between microbial risk assessment and other important microbial food safety concepts like the Hazard Analysis and Critical Control Points (HACCP) and Food Safety Objective (FSO) approaches to food safety; and Modular Process Risk Modeling. This book also reaches beyond microbial risk assessment to cover microbial risk management and communicate about microbial risks in foods. We are also very pleased that this volume includes a case study chapter designed to integrate many of the concepts presented in the rest of the book.

Thanks to Mike Doyle, Series Editor, for asking me to edit this book, and to Eleanor Riemer, ASM Senior Consulting Editor, for shepherding me through the process.

Thanks to all of my friends and colleagues who agreed to give generously of their time in the midst of birth, death, and the responsibilities of their day jobs. This book would not have been possible without all of your efforts.

DONALD W. SCHAFFNER

Microbial Risk Analysis of Foods
Edited by Donald W. Schaffner
© 2008 ASM Press, Washington, D.C.

Qualitative Risk Assessment

Marion Wooldridge

INTRODUCTION TO QUALITATIVE RISK ASSESSMENT

The words "risk assessment" mean exactly that—an assessment of a risk. At its simplest, any method that assesses, or attempts to assess, a risk can be included in this definition. On this basis risk assessments are undertaken regularly by almost the whole population. In the realm of food safety, everyday examples might include assessing the risk of eating packaged food after its labeled "use by" date or assessing the risk of eating unwashed fruit. Clearly, such risk assessments are highly unlikely to employ specific numerical or probabilistic techniques, so they cannot be quantitative risk assessments. But are they *qualitative* risk assessments? The answer depends on what is defined, in the situation being considered, as a qualitative risk assessment. That is, if, within a particular definition accepted by a given body, a qualitative risk assessment requires specific features not present in these "everyday" risk assessments, then they will not be considered as qualitative risk assessments by that particular body.

Both Codex Alimentarius (Codex) (1999) and World Organisation for Animal Health (OIE) (1999) state that qualitative and quantitative risk assessments have equal validity. In this context "validity" means that, if all the requirements of each of the methods are equally well met, then these two methods are equally valid. This implies that, for these bodies at least, specific requirements must be met for acceptance as a valid qualitative risk assessment. The remainder of this chapter describes such requirements, giving guidelines as to how they may be achieved and indicating some of the difficulties encountered and issues that arise.

MARION WOOLDRIDGE, Centre for Epidemiology and Risk Analysis, Veterinary Laboratories Agency (Weybridge), New Haw, Addlestone, Surrey KT15 3NB, United Kingdom.

CHARACTERISTICS OF A QUALITATIVE RISK ASSESSMENT

Main Principles

A risk assessment is the estimate of a risk; a risk is the *probability* of an unwanted outcome, or consequence, occurring. Obviously the possibility for that consequence must exist, but it is not enough simply to prove that possibility (OIE, 1999; Wooldridge, 1999). Therefore, the main principles in a qualitative risk assessment are exactly the same as for a quantitative risk assessment and are summarized in the following text. The remainder of the chapter then discusses issues arising that are specific to qualitative risk assessments.

Use of Qualitative Risk Assessments

A risk assessment gives information, not solutions. The risk manager or policymaker still has the task of deciding whether the assessed risk is acceptable or whether safeguards need to be put in place. Despite several recently published high-profile quantitative microbiological risk assessments, the vast majority of risk assessments utilized over time by risk managers and policymakers in the fields of food safety, health, and microbiology have not been quantitative.

Where a formal risk assessment is commissioned from a risk assessor, reasons for specifying a qualitative risk assessment may include such things as:

- A perception that a qualitative risk assessment is quicker and simpler to complete
- A perception that a qualitative risk assessment will be easier for the (usually nonmathematically trained) risk managers or policymakers themselves to understand
- A perception that a qualitative risk assessment will be easier to explain to third parties
- A personal preference for methods perceived as being more similar to the informal risk assessments that the risk managers may have undertaken for themselves in the past
- An actual or perceived lack of data to the extent that the risk manager believes that a quantitative assessment will be impossible
- A lack of mathematical and/or computational skills and facilities in the risk assessment facilities available or normally used

Whatever the reasons, many of them involve perceptions about formal qualitative risk assessments that are not always true. As can be seen from Table 1, data must be collected, documented, and referenced, whichever method is employed, and this is frequently the most time-consuming part of any risk assessment.

Table 1 The minimum requirements of a risk assessment, summarized[a]

Requirement	Explanation
A. A risk assessment must be transparent. That is, it must be clearly set out and fully referenced in the risk assessment report produced.	
B. The hazard(s) to be addressed must be defined and clearly stated.	In a microbiological risk assessment the hazard is in general the microbe of interest.
C. The risk that the risk assessment evaluates must be defined and clearly set out.	This means specifying the "risk question" to be asked, including defining the outcome(s) (consequences) of interest.
D. The potential pathway(s) (i.e., necessary sequence of events) from the hazard(s) of interest to the outcome(s) of interest must be identified and clearly described. Details of the steps incorporated in this pathway (e.g., testing for the presence of a microbe) must be fully referenced.	These pathways will be based on the (mainly biological) requirements necessary to arrive at the defined outcomes. They are usually most clearly shown in the risk assessment as a series of steps in a diagram.
E. For each identified step in the pathway, information (data) must be gathered to enable evaluation of the probability that that step will occur. It must be clearly set out and fully referenced.	The information (data) will be either qualitative or quantitative, depending on availability. The assessor(s) should undertake a search that is as thorough as is appropriate for the assessment in hand.
F. For each identified step in the pathway, the information (data) is then used to evaluate the probability of that step occurring. Information on uncertainty and variability should be included.	This will be a probability given in either qualitative terms (e.g., high, low, negligible risk or other appropriate text) or quantitative terms. Similarly the information available regarding uncertainty and variability may be narrative or numerical.
G. The overall probability of the complete pathway of events from hazard to defined outcome is evaluated and described as fully as possible.	For a quantitative risk assessment this stage will be computed mathematically. For a qualitative risk assessment this stage may be possible only by summarizing the findings of the individual steps in the risk pathway.

[a]Adapted from a table by Wooldridge in the World Trade Organization (1998), p. 99.

A qualitative risk assessment describes the probability of an unwanted outcome in terms that are by their very nature subjective. This point will be discussed later, but it means that it is not necessarily easier either for the risk manager to understand the conclusions obtained from the risk assessment or to explain them to a third party.

Data are required both for qualitative and quantitative risk assessments. For both, hard numerical data are preferred, and a lack of appropriate crucial data will affect both types of assessments adversely. This point will be discussed more fully later, but if it is impossible to undertake a quantitative risk assessment for this reason, a qualitative risk assessment may be unable to reach more than very uncertain conclusions.

Nevertheless, qualitative risk assessments may be speedier and, although they require an equal degree of logic and considerable numeracy, they do require fewer specialized mathematical and computational resources.

Risk Assessment and Risk Characterization

In both the Codex (1999) and OIE (1999) systems, a risk assessment is the term used to describe the complete process of assessing a risk. In both systems it is broken down into several stages. Looking specifically at Codex, those stages include hazard identification and hazard characterization, exposure assessment, and risk characterization. However, the stage of risk characterization is the process of bringing together all the previous stages to give an overall description, or assessment, of the risk. Thus, the risk characterization step is the process of assessing the risk of interest. The distinction between the two descriptions is therefore subtle. A risk assessment is the process of undertaking the steps that lead to an assessment of the risk, whereas in Codex, the risk characterization is that final assessment of the risk. The same final assessment in the OIE system is called the risk estimate.

With fully quantitative risk assessments, the distinction is relatively simple to appreciate. Appropriate mathematical combination of the estimated probability of exposure, with the estimated dose of pathogen necessary to result in the specified outcome (e.g., infection, clinical illness, death, etc.), will result in the final numerical estimate of the overall risk. This process is the major feature of the risk characterization step.

With qualitative risk assessments, however, there are no numerical exposure assessments to mathematically combine with numerical dose-response estimates; both probabilities will be described in subjective text. In some cases it will be possible to combine these to make logical deductions analogous to mathematical deductions, but in some cases that will be very difficult or impossible (these points will be discussed later). In such cases, the risk characterization step is most likely to be a summarization of the major conclusions from the different stages of the risk assessment. To fully understand the probable magnitude, consequences, and uncertainties of the overall risk, in general, it is necessary to access the complete risk assessment. Thus it may be more difficult to conceptualize the difference between the overall process of a risk assessment and the specific process of risk characterization.

The Complementary Nature of Qualitative and Quantitative Risk Assessments

As shown in Table 1, the main principles of a risk assessment apply equally to both qualitative and quantitative risk assessments. These include identifying the hazard, defining the risk question, delineating the risk pathway, gathering data and information (including that on uncertainty and variability), combining the information in a logical manner, and ensuring that all is fully referenced and transparent. Thus many of the activities are the same up to and including gathering the data. A qualitative risk assessment is therefore often undertaken initially, with the intention of undertaking a quantitative risk assessment only if it is subsequently considered necessary.

However, the qualitative assessment may provide the risk manager with all the information he or she requires. For example, perhaps the data gathered show that the risk is effectively indistinguishable from zero. Or, conversely, perhaps evidence shows that the risk is clearly unacceptably large, or that one or more consequences are so unacceptable that safeguards are needed in any event. A qualitative risk assessment may also provide the necessary insights into the risk pathway(s) that allow the risk manager to make decisions or to apply safeguards without further quantification. In these circumstances the risk manager may decide quantitative assessments are unnecessary.

In some instances a qualitative risk assessment may be undertaken specifically to facilitate a later quantitative assessment. For example, it may be used to identify the data currently available, with attendant uncertainties, to decide whether quantification is currently likely to add value. It may be used to identify areas of data deficiency necessary for quantification and thus target these for future data-gathering studies. It may be used to investigate multiple risk pathways, for example, exposure pathways, prioritizing them for the application of quantification.

Whatever the initial intention, when a qualitative risk assessment has already been undertaken, much of the work for a quantitative risk assessment has been done. For the same risk question, quantification will be able to build on the risk pathway(s) and data collected to provide a numerical assessment of the risk.

Subjective Nature of Textual Conclusions in Qualitative Risk Assessments

It has been indicated on many occasions that assessing probabilities in terms of high, medium, low, negligible, and so on, is subjective, because the risk assessors apply their own concepts of the meanings of these terms. But these meanings will differ, to an unknown extent, from person to person, and this is one of the major criticisms leveled at qualitative risk assessments. However, these assessors' estimates should never be viewed in isolation—which is also, of course, the case with final numerical outputs from quantitative risk assessments, though this is sometimes forgotten.

Within any risk assessment, of either type, judgments will be used throughout. These may be the risk assessors' judgments, or expert opinion, or both, and these will always be subjective. This will apply when selecting (and rejecting) data, delineating the risk pathways, applying weightings to data or model pathways, selecting the distributions used in a stochastic model, as well as selecting a description of high, low, and so on, in a qualitative assessment. Therefore, any risk manager, policymaker, or other stakeholder who needs to use, or wishes to understand, a given risk assessment should not look only at the final "result." They should have some understanding of how that result was reached.

With a quantitative risk assessment, many people may not have the knowledge base to directly understand the computations involved. They will need to rely on the explanations and opinions of the risk assessor to explain how the result was reached and what were the underlying assumptions, judgments, and uncertainties. If the risk assessor is a good communicator this can work. But only then is the risk manager likely to be able to truly understand the significance and meaning of the quantitative result. With a qualitative assessment, providing it has been written in a transparent and logical way, most people should be able to understand and follow the arguments. And even when poorly written, this is likely to be apparent to the risk manager. Therefore, by examining the complete risk assessment, the risk manager (and others) can see directly whether they agree with the conclusions of the risk assessor. Since the usual purpose of a risk assessment is to give a risk manager information to assist with decisions, either of these methods, appropriately carried out, should be equally valid.

But what about the situation where the qualitative description given to a risk manager by the risk assessor (e.g., low risk) is different from that which the risk manager would have used to describe the risk (perhaps, e.g., very low risk), given the same information? Given the way in which risk managers, stakeholders, and others should be using a risk assessment—that is, with an understanding of the whole body of evidence—then it can be seen that this should not actually matter. They will make their decisions on the evidence presented, whatever word they might personally use to describe the magnitude of the risk.

In that case the question arises, is there any point in applying a subjective, descriptive term in a qualitative risk assessment? But despite subjective differences in the meanings of words, some correlation usually exists in the way people use these terms. For example, if 99% of the population were likely to become infected with pathogen P, this would be considered by most people as a "high" risk. Conversely, if potential pathogen P had never been demonstrated to infect humans, despite a high level of exposure and highly sensitive tests, most people would be likely to describe this risk using words such as "very low," "exceedingly low," or possibly "negligible." Thus the description does give a useful broad indication of the magnitude of the risk. This also demonstrates one other use of such descriptions, and that is the possibility of risk prioritization by using them. As long as a single assessor's textual estimates are used consistently within a set of associated risks, this can be very useful for ranking those risks.

It is often suggested that when using terms such as high, low, very low, and so on, in a qualitative risk assessment, a numerical definition of these terms should be given. But unless a quantitative risk assessment has also been done, the actual numerical value of the assessed risk is (by definition) not known, so it cannot be fitted precisely into a numerical categorization system. And if a quantitative risk assessment has been done, then the issue will not arise—the numerical estimate is reported instead. Nevertheless, in some circumstances, it might be requested that risk assessors select values with which they define the numerical boundaries of their own descriptions. If this is done it should be remembered that, if quantified, the risk would not necessarily lie between those numerical boundaries—it is the risk assessor's judgment—and for this reason it may imply unwarranted precision in the qualitative estimate. A sound understanding of the evidence presented, as indicated above, is often more appropriate and useful.

However, there is one case when a definition of text can be particularly useful, and this refers to the description "negligible" to describe the magnitude of assessed risk. One definition of "negligible" used in qualitative risk assessment is that, for all practical purposes, the magnitude of a negligible risk cannot, qualitatively, be differentiated from zero (e.g., see the use of this term in Murray et al. [2004]). The term zero is not actually used because, if a risk is eventually quantified, however small it may be, it will not be zero—thus, the use of the term "negligible," rather than zero.

Qualitative Risk Assessment in Food Safety

Qualitative risk assessments have been used extensively in *import* risk assessments of food products intended for human consumption, especially those of animal origin. However, such import risk assessments have historically been used only to assess the probable *presence* of a pathogen in that product, so that if this probability is unacceptable, import safeguards (e.g., cooking, freezing, quantity control, or total ban) can be applied.

In contrast, food product *human health and safety* risk assessments, in general, set out not only to assess the probability of the presence of a pathogen, but the probable *amount of pathogen* present, in order that the human response to the dose can be assessed. It is this aspect that can make the risk characterization step, and its interpretation, more difficult in qualitative food safety risk assessments—despite the fact that many quantitative dose-response data are very subjective in their estimation methods.

Possibly for this reason, in conjunction with a rapid development in quantitative methods, qualitative risk assessments have recently perhaps been "bypassed" in favor of supposedly more "modern" and objective quantitative methods for food safety and human health risk assessments.

The example in the case study illustrates how qualitative risk assessments can be used in determining the probability of human health effects from food-borne pathogens and food chain exposure, despite lack of data on numbers of organisms present.

Comparison of a Literature Review and a Qualitative Risk Assessment

As has been shown, a qualitative risk assessment is approached initially in a way very similar to that of a quantitative risk assessment. It should be clear, therefore, that a qualitative risk assessment is not merely another name for a literature review. Many other features also differentiate between literature reviews and qualitative risk assessments, and these are described in the next section.

APPROACHES TO QUALITATIVE RISK ASSESSMENT

The Defined Risk Question and Its Importance in Qualitative Risk Assessment

A risk assessment automatically has at its center an issue that is perceived as being "risky," followed by the question, "How risky is it?" To undertake a formal risk assessment, whether qualitative or quantitative, this must be translated into one or more appropriate "risk questions," with hazards and consequences of interest specified.

The specific risk question(s) to be answered are normally the subject of discussions between the stakeholders, in particular, risk managers and risk assessors. The final form of a risk question is usually the result of an iterative process and may change as more information becomes available. Nevertheless, early on in the risk assessment process, an initial risk question should be identified to allow the beginning of the process of answering the question.

One example of such a risk question, taken from a relatively recent qualitative risk assessment (Wooldridge, 1999), is this: What is the risk of adverse human health effects consequent upon the development of antibiotic resistance to (fluoro)quinolones in *Salmonella enterica* serovar Typhimurium that is due specifically to the use of (fluoro)quinolones as veterinary medicines in farm livestock (in a specified region)?

The Risk Pathway and Its Importance in Qualitative Risk Assessments

As described in Table 1, the risk pathway(s) are the potential pathway(s) from the hazard(s) of interest to the outcome (consequences) of interest. That is, they describe the sequence of (mainly biological) events necessary for those consequences to occur. They are, in general, most clearly illustrated as a series of steps in a diagram.

The elucidation and description of such a pathway are as essential for a qualitative risk assessment as for a quantitative assessment, as it is on this model that the format of the whole risk assessment is based. Appropriate data for collection and incorporation are identified, based on the defined steps in the risk pathway. The order in which the data are presented and the identification of the required probabilities and conclusions rely on knowledge of the underpinning steps in the risk pathway.

The need for a clearly presented risk pathway cannot be overestimated, and this will be highlighted throughout this section and illustrated in the case study presented (see "Example case study: antibiotic resistance through the food chain," below).

Data Requirements: a Comparison of Qualitative and Quantitative Risk Assessments

Data used within both qualitative and quantitative risk assessments are likely to include both numerical and narrative information. They fall into two basic categories: the data used to estimate the model input parameters, and the data used to describe the risk pathway and thus construct the model framework. Taking the second type of data first, for both qualitative and quantitative risk assessments these data are usually obtained in the form of narrative text. Discussions and observations on farms or in food-processing plants, for example, will enable a description of the steps in the risk pathway to be elucidated. This description is then usually converted to a diagram, for clarity, and forms the basis of the steps in the model framework.

For a quantitative risk assessment the first type of data, the model input parameters, must *all* be numerical. Where there are no numerical data, methods must be employed to overcome this, for example, use of expert opinion converted to a numerical format. In addition, where uncertainty and/or variability exist, these must be incorporated mathematically, in general, as distributions if a stochastic risk assessment is being undertaken. Where there are several sources of data for a given input parameter, these sources must be weighted and/or combined in appropriate mathematical ways that reflect their importance in estimating the parameter in question.

Despite its name, a qualitative risk assessment still relies, wherever possible, on numerical data to provide model inputs. The search for information, and thus for numerical data, should be no less thorough simply because it is a qualitative assessment. Also, where crucial numerical data deficiencies exist, expert opinion may again be utilized. The major difference lies in how the data and expert opinion are treated once obtained.

Numerical Inputs into Qualitative Risk Assessments

As with quantitative risk assessments, the aim in qualitative risk assessments is to estimate the probability with which each step in the risk pathway will occur. As the probability of risk is most easily estimated with quantitative data, this is used whenever possible and presented in the format in which it is of most use.

To take a very simple example, if a known number of samples have been tested (for microbe M), and a known number found to be positive, then both the actual numbers, plus the percentage positive (for M), should normally be presented. This is because it is much easier to make reasonable inferences about the probability of being M-positive from a percentage than from two separate numbers. This principle of converting raw data to more useful formats using simple numerical manipulations should be utilized wherever practical.

This means that some steps in a risk pathway, even within an overall qualitative risk assessment, may be easily represented by a numerical value, and if this is the case, then this numerical value should be included. However, it should be remembered that such simple estimates are unlikely to be precisely analogous to the estimated parameters in a quantitative risk assessment for a number of reasons. These reasons include, in particular, the way that multiple data sets, uncertainty and variability, and the use of expert opinion are incorporated into quantitative modeling. All these issues are discussed later in the text.

Dealing with Multiple Data Sets: a Comparison of Approaches

Where several studies are available giving data on a particular input parameter, and the value of that parameter varies between the studies, it is not usual in a qualitative risk assessment to attempt to combine them mathematically. Each study and its findings are reported, with an estimate of the importance that should be attached to each, given the particular risk assessment in hand. This has advantages and disadvantages.

Mathematical combination into a single-input parameter makes it possible to estimate with *precision* the probability of that step in the model occurring. However, in either type of risk assessment, the estimate of the importance of each study will be based on the risk assessors' own opinion or that of the selected experts and thus in each case will be susceptible to bias or subjectivity. That is, although *precise*, the parameter in the quantitative assessment may not *accurately* reflect the situation, and this potential subjectivity and bias may (if not transparently described) become hidden in, for example, a weighting applied to the studies within the mathematical model. However, in a qualitative risk assessment, presentation of the individual studies separately means that this particular problem is avoided; it is clear to the reader that there are several possibilities and that the evaluation of their relative importance is subjective and may be biased.

Dealing with Uncertainty and Variability: a Comparison of Approaches

A quantitative risk assessment should take uncertainty (lack of knowledge of the true value of a parameter) and variability (genuine biological or other differences) into account quantitatively in estimating the model input parameters. The specific method may vary and may include the use of distributions, or scenarios (e.g., mean, worst case, etc.). Either way, the uncertainty and variability are reflected numerically in the resulting outputs.

A qualitative risk assessment should also take parameter uncertainty and variability into account. For example, where data giving a range or a specific distribution are available, this should be reported in the risk assessment. However, there is no specific way in which uncertainty and variability in any one input parameter are retained and reflected precisely in the final risk estimate, even when numerical data are available. As with the assessment of risk, the overall assessment of uncertainty and variability from this source will be evaluated in narrative, descriptive terms.

A common source of uncertainty is where the only available data do not relate precisely to the situation in question. Perhaps it is from a different country, or culture, or lacks specificity in its description. Take, for example, a risk assessment where the hazard is microbe species M, subspecies S. Suppose that, universally, data on this microbe are sparse, but there are some data available on microbe M, subspecies unspecified. In a quantitative risk assessment, a decision would have to be made as to whether the range of known subspecies of M was similar enough to S to utilize these unspecified data. Using these data might lead to precision but inaccuracy (if the subspecies were in fact very different), whereas not using these data might lead unnecessarily to a lack of data (if in fact it included subspecies S). The decision would be subjective, based on the risk assessor's or expert's opinions. However, with a qualitative assessment the data can be described as reported, and additional information can be given regarding the similarity (or otherwise) of known subspecies of M. All available data can therefore be utilized and its relevance assessed by any reader, thus avoiding the extremes of discarding or giving too much weight to available data. This should also enhance transparency.

There are other types of uncertainty, however. One is model uncertainty, that is, the precise steps by which the unwanted outcome occurs. In a quantitative model, this may be taken into account by using several different pathways within the model and weighting them (as for multiple data sets) according to the risk assessor's or expert's own opinion. In a qualitative risk assessment the different pathways will be described, ideally with diagrams, and the model uncertainty reported and alternatives discussed. The advantages and disadvantages of the two methods are therefore similar to those described for combining multiple data sets.

The Use of Expert Opinion: a Comparison of Approaches

Where a severe lack of data exists for any crucial aspect of a risk assessment, and it is pragmatically necessary to assess that risk immediately or urgently, in general it is necessary to utilize expert opinion. Problems here include, for example, decisions on identification and selection of experts, the number of experts required, techniques for eliciting information, overcoming bias, and so on, and methods are developing in this area. Whenever expert opinion is required, these problems will be similar for both qualitative and quantitative risk assessments.

It is not the intention here to go into detail on methods involving the use of expert opinion, but it is accepted that ideally a "sufficient number" of experts should be utilized. However, situations exist when there truly are very few experts, perhaps even only one, in the specific topic worldwide. Sometimes there are no true experts. This leads to the use of inputs with very wide levels of uncertainty, whatever the risk assessment type. This is far from ideal but may on occasion be the only option for short-term risk management. Transparency on the implications of this type of uncertainty is more easily made explicit in a qualitative risk assessment.

In a quantitative risk assessment, it is necessary to convert expert opinion into numerical input, and once again various methods exist and are being actively developed (e.g., see Gallagher et al. [2002]). In a qualitative risk assessment, these methods can also be used and may often be the preferred method. However, an alternative way of using expert opinion in qualitative risk assessments is to ask for an opinion on the probability of a specific step in narrative terms of, for example, high, low, negligible, and so on. The meanings of these words will have the same subjectivity problems as has been discussed for qualitative risk assessments in general, and readers' evaluation of the results will need to be based on their evaluation of the experts selected. In theory such a method should be only a temporary measure until improved data are available.

Data Not Relevant to Assessing the Risk in a Qualitative Risk Assessment

Data included in a qualitative risk assessment should contain only that information that is required to assess the risk of concern. The report must not contain irrelevant information, which would only serve to cloud the main issue and make the assessment less transparent.

For example, suppose the issue is the risk to human health from microbe M, due to consumption of product P, due to the presence of microbe M in livestock species S, from which P is produced. The risk assessment will need to include all available data that assist in estimating the probability of the presence and quantity of M for steps on the risk pathway from S, via P, to human consumption and effect. However, information on microbe M that is available but *not* utilized in assessing these probabilities should not be included in the risk assessment report.

Often, when gathering data, more data are collected than are finally needed. In fact, sometimes it is not possible to be sure exactly what might be needed for some steps until the assessment is nearing completion. This leads to a temptation to include all information gathered, relevant or not. This temptation must be avoided, despite frequent pressure from the experts consulted to put in everything *they* know about, for example, a specific pathogen, or disease, or livestock species.

Regarding each piece of information collected, risk assessors should ask themselves: have I needed to use this in coming to my conclusions about the probable level of risk? If not, it should be omitted from the final report.

Order of Data Presentation in a Qualitative Risk Assessment

Two alternative methods for presenting data in a qualitative risk assessment follow.

- A section of the report is allowed for each step in the risk assessment pathway, where relevant data are presented followed immediately by reporting the resulting inferred probability for that step.
- A data section is included where all the data used are presented, categorized as appropriate, followed by an assessment section where the resulting inferred probability for each step in the pathway is described sequentially.

The selection of the presentation format is likely to depend on the complexity of the assessment and whether the same data must be used for multiple steps within the assessment. Whichever method is used, data should be reported in a "logical" order, as far as possible, correlated with the diagram used to clarify that risk pathway.

With the first method of reporting, relevant data should be presented directly at, and for, each step in the risk pathway. Sections or headings within the risk assessment are used to indicate the step being assessed. For example, if the step requires the estimation of the probability of microbe M in livestock species S, then all data that allow that probability to be estimated are placed in a section with a heading that refers to the estimation of that probability. Even within each section, logic should also be used in ordering the data, and this is likely to be similar to the order in which data would be used mathematically if a quantitative risk assessment was being undertaken—and this comparison can act as a good "logic check."

With the second method, often used to avoid excessive repetition of the same data, all data are presented at the start. All data on each relevant aspect are grouped together under the same heading (e.g., "Information on prevalence of microbe M in species S" or "Information on production processes for food product F," etc.), usually with a section number to allow easy later referral. Appropriate data are then referred to by, for example, section number, at each step where it is required in the probability estimates. Even in this format, the data section should be presented in a logical order correlated as closely as possible to that of the risk pathway.

However, each qualitative risk assessment is different in detail, and each will present its own specific issues concerning the most informative and logical way to present the data. An example of one approach is given in the case study (see "Example case study: antibiotic resistance through the food chain" below).

Transparency in Reaching Risk Conclusions in Qualitative Risk Assessments

A qualitative risk assessment should show clearly how each of the risk estimates is reached. This means that the data used to reach any particular conclusion need to be identified in describing the risk estimated and conclusion reached.

The precise way of doing this will vary depending, in part, on the complexity of the risk assessment and, in part, on risk assessor preferences. Methods utilized include a tabular format, with data presented in the left-hand column and the conclusions on risk in the right-hand column, and a format with a summary or conclusions section at the end of each data section. Examples of these formats are illustrated in Tables 2 and 3 for particular steps in the overall risk question: "what is the probability of human illness due to microbe M, in country C, due to the consumption of meat from livestock species S infected with microbe M?" Both examples follow the first format described in the previous section, i.e., presenting data at the relevant

Table 2 Example of a possible tabular format for presenting data linked to risk estimates and conclusions

Step being estimated: What is the probability of a randomly selected example of species S in country C being infected with microbe M?

Data available	Risk estimate and conclusions reached by the risk assessor for this step
The prevalence of microbe M in species S in country C was reported as 35% (Smith and Jones, 1999)	The studies suggest that the probability of a randomly selected example of species S in country Y being infected with microbe M is medium to high. However, the two studies indicate that considerable variability by region is likely. With only two studies available, there is also considerable uncertainty on the actual range of prevalence by region, as well as the probability of infection in a randomly selected sample of species S. In addition, the timing of these surveys may suggest an increasing prevalence of microbe M in country C. The reported parameters for the diagnostic test used do not alter these conclusions.
The prevalence of microbe M in region R, a district within country C, was reported as 86% (Brown, 2001)	
With respect to species S, there are no particular geographical or demographic differences in region R, compared with the rest of country C (Atlas of World Geography, 1995)	
The diagnostic test for microbe M used in the livestock surveillance program in country C is reported to have a sensitivity of 92% and a specificity of 99% (Potter and Porter, 1982)	

Table 3 Example of a possible sectional format for presenting data linked to risk estimates and conclusions

Section X. What is the probability of human ill health, given infection with microbe M?

Data available

No specific dose-response data have been found for microbe M.

Health authorities for country C provide the following data (National Health Reviews, 1999–2002)
- Incidence during the period was reported as 22 cases per million of the population per year.
- Twenty-two cases per million per year equates to 0.000022% of the population per year.

Clinical incidence recording and reporting systems in country C are considered to be of exceptionally high quality (Bloggs, personal communication).

Expert opinion among specialists indicates that once clinical symptoms appear patients are likely to consult a medical practitioner (*Journal of Microbial Medicine*, 1992).

Cases of infection tend to seen in the very young or the very old (*Journal of Microbial Medicine*, 1992).

A surveillance study undertaken by practice-based serological testing indicated that 35% of the population of country C had been exposed to microbe M and had seroconverted (Hunt et al., 2001). This was a countrywide, statistically representational study.

Conclusions

Data suggest a fairly high level of exposure to microbe M in country C, but a very low incidence of clinical disease. Expert opinion indicates that underreporting of clinical disease due to lack of medical practitioner involvement is unlikely to account for this. Overall, therefore, the probability of human ill health, given infection with microbe M, is likely to be low. The level of uncertainty in the data specific to country C appears to be low, making this estimate reasonably certain.

However, data also indicate that specific groups are at higher risk of clinical illness, specifically the very old and very young. From the data currently available it is not possible to indicate how much higher this is likely to be.

step in the risk pathway. If the second format had been used, reference would be made to the section in which the data were to be found, with a brief summary as necessary.

Whatever format is adopted, it is essential to clearly show how the data available are used to reach the risk estimate given. They must be closely and obviously linked.

Risk Characterization: Combining the Steps To Give the Overall Conclusions

This is the stage in qualitative risk assessment that differs most markedly from quantitative risk assessment. For both, the hazard characterization process should have identified the range of possible consequences and their potential severity. For the quantitative risk assessment, for each of those consequences of interest, a mathematical probability will now be computed. Put simply, if the steps in the risk pathway i to n have probabilities x_i to x_n, then multiplying the values from x_i to x_n together will give the estimate of the risk. However, trying to undertake an analogous process with qualitative terms frequently leads to errors of logic. This is because probabilities are defined as being between 0 and 1. Therefore, the highest probability possible is 1, which indicates a certainty. Therefore, even a probability defined as high will in general be less than 1, and at the limit an absolute maximum of 1.

This means that a sequence of probabilities multiplied together must always either remain at the same magnitude (if the next number is 1) or get smaller, and the effect of this is most easily seen in an example. Take a risk assessment with four steps, for each of which a probability has been estimated. Table 4 illustrates two such assessments, one quantitative and one qualitative. However, attempts to "multiply" together qualitative descriptions frequently forget this mathematical analogy and assign medium or even high levels of risk to final

Table 4 A comparison of the computation of the final risk estimate (the risk characterization stage) in quantitative and qualitative risk assessments R1 and R2

Quantitative risk assessment R1[a]		Qualitative risk assessment R2[b]	
Probability	Computation	Probability	Computation
0.1		Low	
0.001	$0.1 \times 0.001 = 0.0001$	Very low	Low × very low → very low or lower
0.5	$0.0001 \times 0.5 = 0.00005$	Medium	... × medium → further reduction
0.9	$0.00005 \times 0.9 = 0.000045$	High	... × high → further (small) reduction
$R1_{final} = 0.1 \times 0.001 \times 0.5 \times 0.9 = 0.000045$		$R2_{final}$ = very low or lower	

[a]Estimated quantitative probabilities from four steps in risk assessment R1, showing how overall risk assessment $R1_{final}$ is computed.

[b]Estimated qualitative probabilities from four steps in risk assessment R2, showing how overall risk assessment $R2_{final}$ is computed.

estimates if these appear as estimated probabilities for intermediate stages. The most that can be said, in this simple case, is that the final risk estimate must be as low as or lower than the lowest individual estimate of probability. Where this "computation" is possible and undertaken logically, it may be useful. However, there are many reasons why it might be impossible to do this.

Frequently with qualitative risk assessments, there are steps with a great deal of uncertainty—often this was the reason for undertaking a qualitative risk assessment in the first place—and it is unlikely that this has been quantified. There may be many steps where such a large amount of uncertainty surrounds the available data that attempting to "compute" a final estimate gives no additional information and becomes meaningless.

There may be model uncertainties, meaning that several possible pathways have been described for part, or all, of the model. There may in fact be several known possible pathways. Where either of these situations is the case, combining qualitative probabilities from these multiple pathways in any logical manner to give an overall estimate is almost certainly going to be impossible.

There is often a step where a probability per random population member has been estimated; e.g., probability of contamination per food item or probability of illness per person in a specific dietary risk group. However, what the risk manager may require is the total probability of the outcome per time period, i.e., taking into account total numbers of servings eaten per year or the number of people who adhere to that specific diet. In this case, the analogous quantitative multipliers are probability (P, range 0 to 1) and integers (I, range 0 to some maximum number). Attempting to "multiply" a qualitative probability with a qualitative estimate of a total number may on occasion lead to a sensible inference (e.g., where the estimated number is so large that a particular outcome is almost inevitable), but in general this is not likely to be the case. And comparison of two or more alternative outcomes, e.g., "low probability per item × many items" versus "high probability per item × few items" in terms of a single descriptor for overall probability may become close to guesswork.

To overcome these problems several methods have been suggested. One method occasionally used has already been outlined in "Subjective nature of textual conclusions in qualitative risk assessments" (see above); that is, risk assessors assign values with which they define the numerical boundaries of their own descriptions. If this is done for each step, taking the mean (or some other value) from these ranges would then allow a computation of overall probability. However, one problem associated with this approach was also indicated in this section, namely, that it is the risk assessor's judgment and that the risk does not necessarily lie between those numerical boundaries. In addition, where there is much uncertainty, this is unlikely to be easily captured in these assigned risk values.

Table 5 An example of the "matrix" approach to risk characterization

Key:
Rows represent the assessed probability per "item" (e.g., probability of exposure to microbe M per soft cheese eater, per year, in country C).
Columns represent the assessed probable number of "items" (e.g., probable number of soft cheese eaters in country C).
Matrix defines the resultant overall probable level of population exposure per year.

	Few	Intermediate	Many
Low	Low	Medium	High
Medium	Low	Medium	High
High	Medium	High	Very high

One frequently considered method involves assigning qualitative inferences to a matrix and arbitrarily assigning overall descriptors to various combinations. Table 5 gives an example of this approach.

Such matrices are often looked on favorably by risk managers and others in the expectation that they will serve to clarify the issues. However, as the result of the different combinations is arbitrarily assigned, there is no certainty that the resulting order of assessed risk magnitude is correct for the whole matrix, and transparency is thereby lost. And crucially, information on the level of uncertainty surrounding each estimate is lost.

Occasionally, a decision tree is considered as a way of summarizing the findings. That is, for each possible risk inference at each step, some specific action is identified. However, this is a risk management tool rather than a risk characterization exercise.

For these and other reasons that become apparent in specific circumstances, therefore, it is frequently undesirable or impossible to "compute" an overall estimate. In such cases, a summary, in a logical order, of the individual conclusions for each step, pathway, etc., including an estimate of the level of uncertainty, becomes the risk characterization process and the information given constitutes a useful précis of the whole risk assessment. This is illustrated in the case study described.

EXAMPLE CASE STUDY: ANTIBIOTIC RESISTANCE THROUGH THE FOOD CHAIN

Introduction

The following example is adapted from a qualitative risk assessment undertaken to assess the risk to human health from the development of antibiotic resistance (Wooldridge, 1999), and is now presented to illustrate many of the preceding points. The example has been generalized to assess the risk to human health from the development of resistance to antibiotic A in *Microbe M strain S* due specifically to the use of antibiotic A as a veterinary medicine in farm livestock. The assessment considers the risk within a defined region of the world, divided into districts each with its own biological and legal characteristics. The example illustrates many of the principles described above in the overall approaches to qualitative risk assessments, and in particular, the following:

- Definition of the risk question
- Description of the risk pathway
- Order of data presentation
- Numerical inputs into qualitative risk assessments
- Dealing with uncertainty and variability
- Data not relevant to assessing the risk
- Transparency in reaching risk conclusions
- Risk characterization: combining the steps to give the overall conclusions

The Risk Question

The specific risk question is: What is the risk of adverse human health effects consequent upon the development of antibiotic resistance to antibiotic A in *Microbe M strain S* which is due specifically to the use of antibiotic A as a veterinary medicine in farm livestock?

The Risk Pathway

At an early stage in the risk assessment, the identified risk pathway was described in tabular stepwise format (rather than the more usual diagrammatic format), and the following steps were identified as shown in Table 6.

Table 6 An example of a risk pathway described in tabular format

Potential risk pathway
Microbe M strain S is present in farm livestock.
Antibiotic A is given.
Resistance develops in a proportion of *Microbe M strain S* due to the use of antibiotic A.
Result: A proportion of the *Microbe M strain S* in livestock is resistant to antibiotic A because of its use in livestock.
Microbe M strain S remains present at all stages in the food chain up to the final product for human consumption.
(and/or *Microbe M strain S* has alternative pathways to ingestion by humans.)
Result: Microbe M strain S (with its resistant proportion) is ingested by humans.
Humans are infected or colonized by *Microbe M strain S*.
Humans become ill because of the *Microbe M strain S*.
Humans ill because of this *Microbe M strain S* are treated with antibiotic A.
Some patients do not respond/respond less well because of resistance.
Result: A proportion of *Microbe M strain S* illnesses treated less successfully than otherwise would be the case because of resistance; that is, an adverse health effect has occurred.

Identification of Data Needs and Format of Report

The probability that these identified steps would occur was then assessed by posing a series of questions. Some questions were further broken down into subquestions. For each question/subquestion, appropriate available data were gathered, wherever possible, to estimate the probability. For each question, reasoned conclusions were then given for the assessment of the probability of that step occurring.

Table 7 gives the list of questions and subquestions posed in the risk assessment, plus other section headings used for data collected in following this risk pathway. Some of the section headings were determined by the type and extent of data available.

The list of questions, and the data collected to answer those questions, is based on the risk pathway outlined in Table 6. Data were thus presented in a logical order, allowing an ordered estimate of the probabilities of the necessary biological steps in that risk pathway, closely analogous to the steps in a quantitative model. Data included numerical data wherever possible, and expert opinion where necessary, including short extracts in the form of concensus statements from reviews, and so on. All were fully referenced. Data not relevant to assessing the probabilities required were not included in the report.

The Risk Assessment Process: One Step in the Pathway

To illustrate the detail within the method, an example of the data available and conclusions reached is given for one section in this risk assessment. The section illustrated concerns the probability of *Microbe M strain S* infecting humans. A description of the data presented in this section of the report is given, followed by the conclusions reached based on the particular data.

Data presented in the risk assessment report

Incidence data from the region of interest were available for two recent years and for the various districts within the region; these data were presented in the risk assessment report (see Table 8). The report also gave background data, and the following relevant points were also included:

- Human infection with this microbe is not reportable in all parts of the region and surveillance methods vary.
- Total *Microbe M* cases recorded are given in the report, plus incidence rates, and for most districts the percentage that were *Microbe M strain S* was also given.

The risk assessment then tabulated these data as follows, making the assumptions described in the table footnotes (for missing data). Note the incidence rate presented here was the recorded rate per 100 humans per year; that is, the

Qualitative Risk Assessment   21

Table 7 An example of questions and subquestions showing the headings used, and thus the order of data presentation, conforming closely to the identified steps in the risk pathway, in this example based on a real qualitative risk assessment

- What is the probability of the presence in farm livestock of *Microbe M strain S* that is antibiotic A resistant due to the use of antibiotic A in those animals?
  - What is the probability of *Microbe M strain S* being present in farm animals in the region of interest?
    - Information on the prevalence of *Microbe M strain S* in farm livestock in the region
  - Can the use of antibiotic A cause resistance per se?
  - What is the probability of antibiotic A resistance, in *Microbe M strain S*, in farm livestock (in the region), and what evidence do we have that it developed because of the use of antibiotic A?
    - Information on the use of antibiotic A in farm animals in the region
    - Information on antibiotic A resistance in *Microbe M strain S* isolated from farm livestock in the region
- What is the probability of human exposure to antibiotic A-resistant *Microbe M strain S* that originated in farm livestock?
  - Direct food chain route for the passage of *Microbe M strain S* from farm livestock to human exposure
    - Information on the possibility of *Microbe M strain S* passing through the direct food chain from animal to human
    - Information on the probability of food of animal origin being infected/contaminated with *Microbe M strain S* from that animal origin
    - Information on the probability of any *Microbe M strain S* found in the direct food chain being antibiotic A resistant
  - Routes other than via the direct food chain by which *Microbe M strain S* may pass from livestock to result in human exposure
  - What non-livestock-originating sources of *Microbe M strain S* are there that may potentially contaminate the food chain, environment, or any other routes of human exposure to *Microbe M strain S*, and/or what alternative sites of development of antibiotic resistance in *Microbe M strain S* are there other than in farm animals?
- What is the probability of *Microbe M strain S* from animal products in the food chain colonizing/infecting humans?
  - The probability of ingesting a *Microbe M strain S* contaminated item of foodstuff of animal origin, and the probability that it contains an infectious dose
    - Information on the probability of ingestion
    - Information on the probability of carriage of an infectious dose
  - The probability that *Microbe M strain S* strains isolated from humans originated from farm livestock
  - The probability of *Microbe M strain S* infecting humans
- What is the probability of adverse health effects in humans due to the presence of antibiotic A-resistant *Microbe M strain S* derived initially from farm livestock due to the use of antibiotic A in those farm livestock?
  - The probability of resistance to antibiotic A in human *Microbe M strain S* isolates
  - The probability of human cases of *Microbe M strain S* requiring treatment with antibiotic A

Table 8 Recorded infection with *Microbe M strain S* in humans in region R (by district D) for year Y1 (or year Y2*): incidence rate per 100 humans per year

District	No. reports	Incidence rate	District	No. reports	Incidence rate
D1	502	0.006	D10	No data given	–
D2	3,750	0.037*	D11	205	0.002^a
D3	838	0.016	D12	620	0.002
D4	399	0.008	D13	459	0.005
D5	6,392	0.011*	D15	1,011	0.007
D6	45,103	0.055*^{b}	D16	826	0.014
D7	10	0.008*^{b}	D17	172	0.010
D8	315	0.010	D18	6,686	0.013*
D9	14,751	0.021^a			

^{a} Percentage figures by serotype not given. This estimate assumes 100% are *Microbe M strain S*.
^{b} Percentage figure given only for *Microbe M strain O*. This estimate assumes ALL others are *Microbe M strain S*.

percentage of the district's population recorded as a case of *Microbe M strain S* infection per year.

The additional point was then made: if testing is assumed to be highly specific, the above estimates represent a minimum; not all *Microbe M strain S* infections will be recorded and reported.

The conclusions reached in the risk assessment

The conclusions reached, based on the data available at the time and as described above, were as follows:

> The probability of any randomly selected human from any district of the region being reported to the appropriate authorities as a case of *Microbe M strain S* is low, and in some districts very low. However, it is highly unlikely that all cases seen by medical practitioners will be reported further, and this depends on the local reporting system. Taking into account the survey method used, in conjunction with expert opinion, uncertainty limits could be constructed for these incidence values, by district.

In fact, this particular risk assessment also gave recommendations (as required in the original remit). One such recommendation was that appropriate information should be further sought and an assessment of the uncertainty limits should be made.

As seen from this example, the conclusion describes the risk assessor's estimate of the probability of that particular event occurring and the descriptions used—*low* and *very low*—to represent the incidence numbers are subjective. However, because the data are also available, any other person reading the report can immediately see on what numbers, and assumptions, these subjective descriptions are based.

Risk Characterization: the Overall Conclusions

Following the completion of the detailed sections corresponding to the risk pathway, as described above, an overall risk assessment summary was given in tabular format. This is essentially the final risk characterization process. Table 9 is based on this.

Points to note in the summary above are that, wherever possible, guidance on the typical values or ranges of magnitude found in the numerical values reported earlier is included along with a textual description of the probabilities. This means that the reader does have an appreciation of what the actual values are, and even if reading only this summary, is not totally reliant on the subjective textual description by the risk assessor. Information on variability and uncertainty is given, along with any other key features for the section under consideration.

This example has illustrated at least one method of dealing with most of the points described in the approaches to qualitative risk assessment.

Points Not Illustrated within This Case Study

Some important points discussed in the text but not illustrated by this example include:

- A diagrammatic description of the risk pathway; although this is a very common and useful feature, it was replaced by text (as shown in Table 6) in this particular assessment
- The alternative order of data presentation described in "Order of data presentation in a qualitative risk assessment" (see above)
- Methods for the use and incorporation of expert opinion
- The use that can be made of simple calculations, even within a qualitative risk assessment

SUMMARY

Formal qualitative risk assessments follow the same basic principles as quantitative risk assessments. They require a specified risk question, and they rely on a model—the risk pathway—to underpin the format of the assessment. The search for numerical data to input into the assessment should be equally as thorough as for quantitative risk assessments. They must address the probability of an unwanted event, not just the possibility.

Specific issues concerned with qualitative risk assessments discussed in the chapter include the reasons for use of a qualitative risk assessment, the differences between a complete qualitative risk assessment and the specific step of qualitative risk characterization (sometimes less distinct than in a quantitative risk assessment), and the complementary nature of qualitative and

Table 9 An example of the risk characterization stage (or risk assessment summary) of a qualitative risk assessment

Key stage of assessment	Conclusion
What is the probability of the presence in farm livestock of *Microbe M strain S* that is antibiotic A resistant due to the use of antibiotic A in those animals?	Information available indicates variation in prevalence of *Microbe M* spp., including *Microbe M strain S*, both between districts within the region and between livestock species. Overall, the data indicate low prevalence of *Microbe M* infection in general. However, much information crucial to proper interpretation of the data is missing, leading to a high level of uncertainty. Values reported for *Microbe M strain S* range from 0 to 9%. Information on the probable presence of resistance and its association with use of antibiotic A is complex and contradictory; the data currently supplied do not allow this relationship to be elucidated. In addition, there is a variation between district and species, possibly associated with usage. For a randomly selected isolate of *Microbe M strain S* there appears to be a fairly low probability that it is resistant to antibiotic A, many data sets giving 0%. However, the range demonstrated (0 to 86%), small numbers, and missing information give a very high level of uncertainty surrounding that low probability. If it is assumed that no inherent resistance exists, then a low proportion resistant within a variable but low prevalence of infection gives a low overall probability of resistant *Microbe M strain S* due to use of antibiotic A. However, due to data deficiencies this is associated with a high degree of uncertainty, in addition to the species and district variability.
What is the probability of human exposure to antibiotic A-resistant *Microbe M strain S* that originated in farm livestock?	There are several routes by which bacteria from animals might pass to humans. This assessment deals mainly with the food chain route. Food contamination with *Microbe M strain S* (or any other strain of *Microbe M*) does not necessarily mean that it came from the source animal. Other sources, including contamination from human food handlers, are possible. Because of data deficiencies, it is not possible to determine what proportion did come directly through the food chain; the maximum (i.e., the conservative—and probable overestimate) is 100% of that present. The probability of the isolation of *Microbe M* spp. from the food chain depends on the stage in the production chain, species, and country; many values are below 1%, although for poultry higher values up to 45 to 50% are reported. Later stages in the processing chain probably have a higher probability of contamination than earlier ones. Specifically for *Microbe M strain S* the probability is lower than for *Microbe M* in general, but serotype is frequently missing. However, for many of the higher values (cattle, poultry) textual information indicates many of the isolates were probably not *Microbe M strain S*. Because of data deficiencies there are wide uncertainty limits on these estimates.

Table 9 (continued)

Key stage of assessment	Conclusion
	The probability of cooking and its effect on the prevalence of viable *Microbe M strain S* in food of animal origin at the point of ingestion has not been adequately addressed. It is likely that this will significantly reduce the probability of ingestion of viable *Microbe M strain S*; where heat treatment was recorded prevalence of *Microbe M* spp. was below 0.5%.
	Overall, it is likely that the probability of a food item of animal origin at the point of ingestion by a human being contaminated with *Microbe M strain S* is very low, whatever the source of contamination, although there is variation within species and country, and a high level of uncertainty surrounding that estimate.
What is the probability of *Microbe M strain S* from animal products in the food chain colonizing/infecting humans?	Conclusions reported are equivocal on the relationship between the isolates of *Microbe M strain S* reported in humans and those in farm livestock. From the data available, it cannot be automatically assumed that all (or even most) of *Microbe M strain S* strains isolated in humans have originated unchanged from farm livestock.
	The probability of any randomly selected human from any district in the region being reported to the appropriate authorities as a case of *Microbe M strain S* infection (from whatever source) is low, and in some districts very low, although there is considerable variation by district; incidence rates per 100 humans per year varies from 0.002 to 0.055. The degree of uncertainty with which this represents the level of illness due to *Microbe M strain S* also varies by district, depending on the reporting system. However, even in districts with mandatory reporting (and thus less uncertainty) the probability is still low.
What is the probability of adverse health effects in humans due to the presence of antibiotic A-resistant *Microbe M strain S*?	The probability of a human isolate of *Microbe M strain S* from the region being resistant to antibiotic A varies with district; in general, prevalence values are low (about 1%), but in one district there is a suggestion that it has increased over time to about 12%. Further, more recent information is necessary to enable an estimate of the degree of uncertainty for these values.
	For *Microbe M* spp., in general, the probability of illness requiring specific antibiotic therapy appears to be low, perhaps up to 5%. For resistant strains that probability may be increased, possibly more than doubled, although the range suggested (10 to 36%) indicates wide uncertainty limits.
What is the probability of adverse health effects in humans due to the presence of antibiotic A-resistant *Microbe M strain S* derived initially from farm livestock due to the use of antibiotic A in farm livestock?	At each stage of the assessment the information suggests only a low probability of the occurrence of the necessary step in the chain of events leading to the unwanted outcome. For most stages some data are available to quantify the level. However, for certain steps data are particularly sparse, in particular, the probability of the food at the point of ingestion being contaminated with *Microbe M strain S* (from any source, and therefore also from farm livestock) and the probability of

(continued)

Table 9 *(continued)*

Key stage of assessment	Conclusion
	antibiotic A-resistant *Microbe M strain S* strains isolated from humans being the same as those strains that are isolated from farm livestock.
	At all steps in this assessment there is variability due to species and district. At those steps where numerical data exist, in general, very wide uncertainty limits exist associated with those numerical data because of lack of critical information on, for example, serotypes, denominators, and survey or reporting methods.
	Therefore, based on currently available information, the probability of adverse health effects in humans due to the presence of antibiotic A-resistant *Microbe M strain S* derived initially from farm livestock due to the use of antibiotic A in those farm livestock appears to be low overall, but with a high degree of uncertainty in this estimate and with much variation by district and species of livestock. Further information might alter this conclusion.[a]

[a] Note that this risk assessment was undertaken in 1998. Further data might now be available to alter those conclusions.

quantitative risk assessments. A qualitative risk assessment will provide a very good basis for undertaking a quantitative risk assessment, if required, as much of the work is already done.

Qualitative risk assessments are sometimes criticized as being "subjective" and "just a literature review." However, both quantitative and qualitative risk assessments are subjective, probably equally so but in different ways, and this point is discussed and demonstrated. This is because a qualitative risk assessment can, because of its narrative nature, actually give far more information in an insightful manner, leading possibly to greater transparency and less hidden subjectivity than a quantitative assessment. The difference between a qualitative risk assessment and a typical literature review should be clear, both from the approaches described and the case study used for illustration.

Specifically, qualitative risk assessments should present the information used to assess the risk in a logical and transparent manner, based on the risk pathway, and describing the variability and uncertainties surrounding that information. The assessed risks or conclusions for each step in the pathway should be clearly related to the data on which they are based. All data and experts' or assessors' opinions should be fully referenced, and all assumptions stated, for transparency. Data not used or necessary in assessing the risk(s) of interest should not be included in the risk assessment, because those data cloud the issue and reduce transparency. Wherever possible, data should be numerical, and simple mathematical calculations to maximize the usefulness of the data are encouraged; they do not turn the process into a quantitative risk assessment (which by definition has a final numerical output).

Risk characterization is the stage in qualitative risk assessment that differs most markedly from quantitative risk assessment. It is sometimes possible to arrive at a narrative, textual, overall estimate of the risk, based on a mathematically logical combination of stepwise assessed risks. However, attempting this can easily lead to errors of probability logic, and in any situation when there is much uncertainty, or multiple pathways, it is likely to be impossible. Therefore, in general, it is more appropriate and useful to provide a summary, in a logical order, of the individual conclusions for each step, pathway, and so on, including an estimate of the level of uncertainty. This then becomes the risk characterization process and the information given constitutes a useful précis of the whole risk assessment. These points are all illustrated in the case study described.

A qualitative risk assessment will provide a very good basis for undertaking a quantitative risk assessment, if later required, as much of the work is already done, and each (given that they are properly undertaken) is an equally valid method. Therefore, the choice between a qualitative risk assessment and a quantitative risk assessment is often not initially necessary. It is frequently useful to first undertake a qualitative assessment, and a quantitative one can then follow, building on the qualitative model and data, if considered necessary.

As stated earlier, a risk assessment is intended to provide one of the sources of information for a risk manager or policymaker to use in deciding whether a risk is acceptable. The numerical result from a quantitative risk assessment, per se, does not necessarily make it any easier to decide whether a risk is acceptable.

ACKNOWLEDGMENTS

Much of this chapter is based on an original paper entitled "Qualitative Risk Assessment." The original paper was prepared for use in specific sections of the World Health Organisation/Food and Agricultural Organisation (WHO/FAO) Guidelines on Risk Characterisation of Microbiological Hazards in Food. It is used here with the kind permission of FAO and WHO. The European Medicines Evaluation Agency (EMEA) is also acknowledged for their support for work from which some examples in this chapter are adapted.

REFERENCES

Codex Alimentarius (Codex). 1999. *Principles and Guidelines for the Conduct of a Microbiological Risk Assessment.* Codex Alimentarius Commission CAC/GL-30. Food and Agriculture Organization of the United Nations, Rome, Italy.

Gallagher, E., J. Ryan, L. Kelly, Y. Leforban, and M. Wooldridge. 2002. Estimating the risk of importation of foot-and-mouth disease into Europe. *Vet. Rec.* **150**:769–772.

Murray, N., S. C. MacDiarmid, M. Wooldridge, B. Gummow, R. S. Morley, S. E. Weber, A. Giovannini, and W. Wilson. 2004. *OIE Handbook on Import Risk Analysis for Animals and Animal Products*, vol. 1. *Introduction and Qualitative Risk Analysis*. World Organisation for Animal Health (OIE), Paris, France.

Wooldridge, M. 1999. Qualitative risk assessment for antibiotic resistance. Case study: *Salmonella typhimurium* and the quinolone/fluoroquinolone class of antimicrobials. In *Antibiotic Resistance in the European Union Associated with Therapeutic Use of Veterinary Medicines: Report and Qualitative Risk Assessment by the Committee for Veterinary Medicinal Products*. European Medicines Evaluation Agency (EMEA), London, United Kingdom.

World Organisation for Animal Health (OIE). 1999. Import risk analysis, chap. 4. In *OIE Animal Health Code*. World Organisation for Animal Health (OIE), Paris, France.

World Trade Organization (WTO). 1998. *Australia—Measures Affecting Importation of Salmon: Report of the Panel*. World Trade Organization WT/DS18/R 12 June 1998. Geneva, Switzerland.

Using Risk Assessment Principles in an Emerging Paradigm for Controlling the Microbial Safety of Foods

Richard C. Whiting and Robert L. Buchanan

INTRODUCTION

Instituting controls over pathogenic microorganisms in foods reaches back to the beginnings of civilization with the development of salting, fermenting, drying, and cooking processes. In fact, it can be argued that the ability to preserve foods for an extended period was a prerequisite to the emergence of large urban cultures. Oral traditions and, later, written instructions codified the empirical knowledge of proper and safe food manufacture. With emerging understanding of disease processes and public health and the rise of nonlocal food processing, public health regulations were passed that focused on sanitation and an absence of filth. An early performance criterion in the United States was the 1939 Milk Ordinance and Code that specified time and temperature requirements for pasteurizing milk (Institute of Medicine/ National Research Council [IOM/NRC], 2003). Following discovery of pathogenic microorganisms and the role of foods in facilitating their growth and distribution, specific requirements for processes were promulgated and enforced by government agencies. These requirements remain a major part of current food regulations and delineate specific formulation or process criteria that must be met. Table 1 shows examples of these process criteria intended to ensure food safety.

This approach to setting process controls has served industry and consumers well for many years. Process and product criteria of this type are readily disseminated, applicable to all processors, specific, and relatively uncomplicated for industry to follow and regulatory agencies to verify. However, this approach is becoming increasingly restrictive. It does not allow

RICHARD C. WHITING AND ROBERT L. BUCHANAN, Center for Food Safety and Applied Nutrition, Food and Drug Administration, 5100 Paint Branch Parkway, College Park, MD 20740-3835.

Table 1 Traditional process criteria for food safety

Product	Pathogen	Process criteria
Fluid milk	*Coxiella burnetti*	72°C for 15 s
Liquid whole eggs	*Salmonella*	149°F for 3.5 min
Poultry meat	*Salmonella*	Minimum 165°F
Shellfish	Parasites	Frozen < −35°C for 168 h
Retail food code	*L. monocytogenes*	a_w < 0.95 and pH < 5.5
Low-acid canned foods	*C. botulinum*	12 D inactivation; 121°C for 3–6 min

processors to alter the process, is not conducive to process innovation or alternative technologies, considers the single process step in isolation from the complete farm-to-fork chain, and cannot be directly linked to public health. High-pressure processing and pulsed electrical field pasteurization have many nontraditional parameters that must be controlled in these recently developed processes to achieve the desired inactivation. For example, with pulsed electrical fields the geometry of the inactivation cell is an important design criterion. Combination processes are also being developed where the temperature is increased during high-pressure processing or when liquid eggs are heated by plate pasteurizers and electromagnetic energy. To achieve appropriate pasteurization with these processes, careful collaboration between microbiologist, food technologist, and design engineer will be needed for each specific combination of food, equipment, and pathogen. It would be impossible for regulatory agencies to specify traditional process criteria for these systems.

Rather than promulgating specific process criteria, regulatory agencies have begun to specify the microbiological results, i.e., performance criteria, to be achieved by pasteurization processes. The FDA has required, for example, that processors of apple juice demonstrate that a 5 $\log_{10}$ inactivation of *Escherichia coli* O157:H7 is achieved. Furthermore, this inactivation need not occur during a single step but may be the sum of several validated steps under a processor's control. The Food Safety and Inspection Service (FSIS) requires a 7.0 $\log_{10}$ reduction in *Salmonella* in cooked, ready-to-eat poultry products. Specific time-temperatures are not given and manufacturers can choose the process that best suits their product, provided they can validate that this reduction is being met.

HAZARD ANALYSIS, CRITICAL CONTROL POINTS

Hazard Analysis, Critical Control Points (HACCP) evolved as a system to control the threats (risks) from food-borne hazards (Buchanan and Whiting, 1998; IOM/NRC, 2003). This preventive system of seven steps requires that processors analyze their product and process for anticipated hazards, identify and validate measures to control these hazards, specify management actions to monitor and verify control has been achieved, and respond to deviations from the HACCP plan. The seven steps are:

1. Hazard analysis
2. Identification of critical control points (CCPs)
3. Establishment of critical control limits for each CCP
4. Establishment of monitoring procedures for each CCP
5. Establishment of corrective actions
6. Establishment of record-keeping procedures
7. Establishment of verification procedures.

The goal of the HACCP approach is to prevent illnesses by designing and controlling the process, in contrast to previous quality control/quality assurance programs that attempted to catch potentially hazardous products after they were manufactured. However, the application of HACCP principles still does not lead to a comprehensive design of a food process. Each process step that is deemed "critical" will have process criteria assigned to it. There is no mechanism in the traditional HACCP system to link the different steps to each other, to quantitatively determine how much control is necessary, or to determine the impact of a critical step on public health and the incidence of food-borne disease.

FOOD SAFETY OBJECTIVE

The food safety objective (FSO) is the number of food-borne pathogens that an individual could consume in a serving that meets the safety criteria (see below for further definition of the FSO). The conceptual "food safety objective" equation was proposed to overcome some of the limitations of the HACCP plan (International Commission on Microbiological Specifications for Foods [ICMSF], 2002; IOM/NRC, 2003). It was recognized that the numbers of pathogens at consumption depended on the initial contamination, the extent of inactivation during the process, and growth by survivors or reintroduction of a pathogen before consumption. This concept was expressed by the relationship:

$$H_0 - \Sigma R + \Sigma I \leq \text{FSO}$$

The maximum number of pathogens at consumption must be less than or equal to the initial level of contamination (H_0), less the summed decreases from all of the steps that remove or inactivate the pathogen (ΣR), and plus the summed increases by growth or recontamination (Σ I). This concept recognized the interrelationship of all the processing steps (including storage periods and home/food service food handling) and that the numbers of pathogens at consumption were related to and a measure of public health. The FSO would be openly and transparently set by an appropriate regulatory agency with consideration of the epidemiology and virulence of a specific pathogen and the susceptibility of different human populations. Food processors would then be responsible and permitted to design and control the process as they preferred, so long as it met the FSO.

The impact of each of the specific process steps, as determined by their CCPs, can be related to their respective contribution to the final number of pathogens at consumption. The FSO concept, therefore, overcomes several limitations of the HACCP plan. However, the development of risk analysis techniques in other fields of science and engineering has shown limitations in the FSO concept. The principle limitation is the lack of consideration in the inherent variation in a food process chain and the realization that evaluation of the "average" process is not a satisfactory measure of process safety. A more complete mathematical treatment of a food process is necessary.

MICROBIOLOGICAL MODELING AND RISK ASSESSMENT

Microbiological modeling is based on the premise that changes in microbiological numbers can be experimentally measured and mathematical equations can be fitted to the data that describe these changes (Whiting and Buchanan, 2001; McKellar and Lu, 2004; Ross, 2008 [see chapter 3]). Each step in the food-processing chain is viewed as a "unit operation." An initial number of microorganisms enter the step and they multiply, decrease, or stay constant. What happens depends on the microorganism, length of time in the step, and environment (temperature, pH, water activity, etc.). An appropriate model is applied for each step where entry of the environmental parameters leads to the calculation of the expected microbial numbers at the end of the step (Whiting and Buchanan, 2001; Nauta, 2008 [see chapter 4]). This new population is passed on to the next processing step.

Log-linear models for the thermal inactivation of microorganisms, i.e., D- and z-values, have a lengthy history with their application in low-acid thermal processing beginning in the 1920s. Nonlinear declines are observed for thermal and other inactivation processes; logistic and Weibull models are examples of mathematical functions that are utilized to describe these declines. In recent years, models were developed for microbial growth of most food-borne pathogens in a variety of broths and foods. The three most frequently used growth models are the simple 2- or 3-phase lag/log-linear, the

Gompertz, and the Baranyi models. In addition, survival models for pathogens in foods that do not permit growth (high acidity or salt) and inactivation models for nonthermal processes have been devcloped. Simple shoulder/log-linear declines, logistic, and Weibull functions can describe these data.

For growth models, as well as survival and inactivation models, research is showing that the presence and length of the lag period are greatly affected by the physiological state of the cells at the beginning of the step (Whiting and Buchanan, 2006). Examples of different physiological states that can influence the lag phase are exponential versus stationary phase cells, heat-shocked cells, starvation, desiccation, acid adaptation, and injured cells. This means that to predict the lag phase, information must be known about the previous environment and history of the cells. Therefore, to accurately model the steps in a food-processing system that consists of a series of unit operations, knowledge of the prior steps has to be considered to accurately estimate lag times. Appropriate individual models can then be linked from raw ingredients through processing, storage, and consumption to estimate the final numbers of a pathogen that would be present.

Models have also been developed to estimate the health impact of consuming specific numbers of microbial pathogens. These dose-response relationships consider the pathogenicity of the microorganism (virulence) and the susceptibility of the person consuming the pathogen. The type of model that effectively describes a dose-response relationship is dependent, in part, on the microorganism's mechanism of pathogenicity. Microorganisms may produce a toxin in the food before consumption (toxigenic: *Staphylococcus aureus*, *Clostridium botulinum*), colonize and produce toxin while in the gastrointestinal (GI) tract (toxicoinfectious: *Clostridium perfringens*, *E. coli* O157:H7), or invade the cells of the GI tract and other organs (infectious: *Listeria monocytogenes*, *Salmonella*). Different strains of a pathogen have different pathogenicities depending on the specific virulence factors that they possess in their genome. Human susceptibility varies among different populations. Many microorganisms are opportunistic and affect individuals with weak or suppressed immune systems. Children, the elderly, pregnant women, patients taking immunosuppressing drugs, or individuals with immune system diseases are typically more vulnerable than the general population, in particular, to infective pathogens. Conversely, individuals can develop significant immunity against certain food-borne pathogens. For example, previous exposures to hepatitis type A virus provide immunity against subsequent exposures. This also appears to be important in some parasitic diseases such as *Cyclospora cayetanensis*.

The environment that the cells inhabit prior to consumption can also affect expression of virulence factors and alter the dose-response relationship. Commonly referred to as food matrix effects, different environments can stimulate expression of virulence factors by pathogens (e.g., stimulate the expression of adhesion factors) and/or increase the likelihood that they reach the site of infection. Microorganisms that are acid adapted or are entrapped in a high-lipid matrix may survive passage through the stomach and cause illness with lower numbers than when the pathogens are in another matrix. For example, the acid adaptation of pathogenic *Vibrio* species can decrease the ID_{50} by several orders of magnitude.

Various functions have been used to describe the dose-response relationship; the exponential, beta-Poisson, and Weibull models are most frequently used (Holcomb et al., 1999). Guidance on "best practices" for the selection of dose-response models for infectious and toxicoinfectious food-borne pathogens has been developed by the Food and Agriculture Organization and the World Health Organization (FAO/WHO) (2003). Because of the wide variation in both the individual susceptibilities of humans (i.e., highly susceptible to effectively immune) and the virulence of individual strains of a pathogen, it is not surprising that the variability associated with a dose-response relationship is substantial. However, this does not preclude its use in estimating the public health consequence of exposures to food-borne pathogens.

It is currently feasible to model, with varying degrees of certainty, an entire food process, linking through consideration of the microbial quality of the ingredients along with the effects of processing treatments that affect microbial growth or survival during manufacture, distribution, marketing, preparation, and consumption of the food to estimate consumers' exposures to a pathogenic microorganism. Such exposure estimates can be used in combination with information on the virulence of the pathogen to estimate the impact on public health (e.g., the probability of illness per serving). If the number of servings per year by a population is considered, the estimate can be made of the total number of cases in that population per year. When combined in this manner, these models constitute a risk assessment. The estimation of the numbers of a pathogen consumed is, in risk assessment terminology, the exposure assessment; the modeling of the dose-response relationship is the hazard characterization; and linking these two to provide a risk estimate is the risk characterization (Whiting and Buchanan, 2001).

FSO/PO PARADIGM

A microbial risk assessment model can be employed to estimate the numbers of pathogens and/or their frequency of occurrence at individual processing steps from raw ingredients to consumption and relate the corresponding exposure to the likelihood of illness or death. These numbers and frequencies may be very low but they are not zero; therefore, quantitative objectives must be formulated. An effectively designed risk assessment can be used to help determine whether a process meets specified public health standards for illness and can estimate the impact that changes in the process would have on the likelihood of illness. Internationally, HACCP is the risk management system employed by the food industry to establish and maintain the designated conditions (CCPs) to achieve the public health standards. To integrate HACCP and risk assessment concepts, a series of terms have been defined (Codex Alimentarius Commission [CAC], 1997, 2004; FAO/WHO, 2002). One of the easiest ways to introduce these terms is to do so in relation to the manufacture and consumption of a food. Figure 1 illustrates a generic food process where an individual serving starts as raw ingredients with some frequency and level of contamination, is pasteurized to reduce pathogen numbers, stored under conditions where survivors may grow, and consumed where the food may cause illness. The FSO terms and their points of application in the food process are indicated.

The paradigm commences with the appropriate level of protection (ALOP), a concept established by the Sanitary and Phytosanitary (SPS) Agreement of the World Trade Organization. This is the agreed upon public health goal or level of protection (LOP) for the food (see definitions below) that the process is then designed to meet. The dose-response relationship for the selected susceptible human population will determine the number of pathogens consumed with a serving that has a risk equal to or below the ALOP. This number of pathogens is the FSO. Factoring in the serving size will convert the FSO to a per gram basis. The maximum numbers of

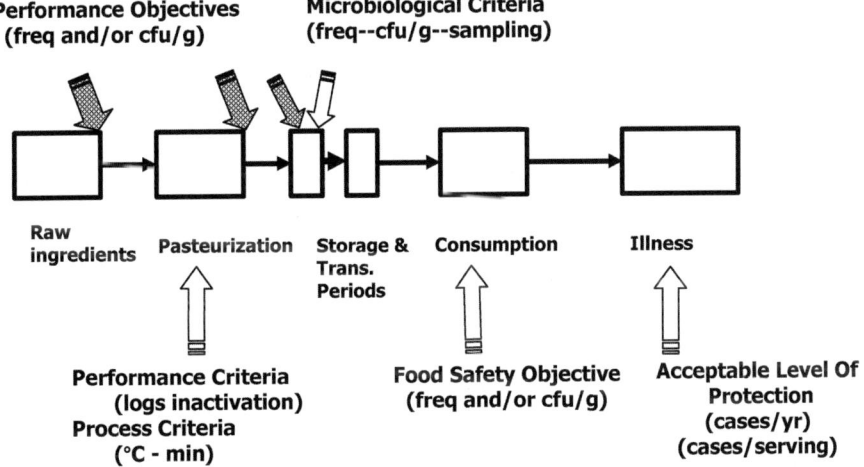

Figure 1 Food safety terms and their point of application in the food process.

the pathogen at earlier points in the processing chain are indicated by the performance objectives (POs); there may be more than one in a food process. The FSO can be viewed as the PO at consumption. If the food supports growth, allowance for growth between manufacture/retail and consumption is necessary. This requires designation of home and food service storage temperatures and times. This would typically include consideration of a reasonable level of abuse. However, the degree of abuse included in defining PO must be pragmatic; most foods that allow pathogen growth would ultimately have counts exceeding the FSO if held for extended periods under overtly abusive conditions. If a process step inactivates the pathogen, the decrease that should be achieved is termed the performance criterion and would be expressed as the log decrease. Performance criteria could also be established for a storage period where a 1 log increase, for example, would be the maximum increase that would still allow the FSO to be met. To reduce microbial numbers, there are many inactivation processes that could achieve a specified performance criterion. Thermal inactivation is the traditional process but microwave, high pressure, or pulsed electrical fields are examples of other inactivation processes that might be employed. There are usually multiple combinations of parameters for each process that could achieve the performance criteria, e.g., different time-temperature combinations that achieve 7.0 logs of inactivation. The specific operating parameters chosen to achieve the performance criteria are the process criteria. Typically, these parameters become the critical control points for the process in the HACCP system.

The FSO and PO are defined in terms of maximum frequencies and/or concentrations, point values or "bright lines" that the food should not exceed. However, to be useful for regulatory and trade purposes these definitions need to be interpreted in terms of the degree of confidence that will be required to ensure that an FSO or PO is not being exceeded. Traditionally, this is achieved by establishing a sampling plan (i.e., microbiological criterion [MC]) against which individual lots of food can be evaluated to determine whether they meet the designated PO. The most commonly employed MCs are attribute (presence/absence) testing plans that include the PO, confidence level at which the sampling plan is capable of rejecting lots that are unacceptable (i.e., have servings that exceed the PO), number of samples taken, number of samples that can be positive, and a sampling-testing protocol. Because a relatively high confidence level of rejecting a lot that exceeds the PO is usually desired, frequently 95%, the log CFU/g value for the mean of the lot will be lower than the PO. A lot with high between-sample variation will need to have a lower mean than a lot with low variation.

The interactions of these terms are illustrated for a simple hypothetical example of POs and an MC for a psychrotrophic pathogen capable of growing

Table 2 Illustrative values for food safety parameters for fresh-cut lettuce and a hypothetical microbial pathogen

Level of protection log (illness/serving)	FSO log (CFU/serving)	FSO log (CFU/g)	PO/MC retail log (CFU/g)	PO/MC manufacturing log (CFU/g)	PO raw lettuce log (CFU/g)
−4.0	6.0	4.3	2.3/1.3	1.3/0.3	1.2
−5.0	5.0	3.3	1.3/0.3	0.3/−0.7	0.2
−6.0	4.0	2.3	0.3/−0.7	−0.7/−1.7	−0.8
−7.0	3.0	1.3	−0.7/−1.7	−1.7/−2.7	−1.8
−8.0	2.0	0.3	−1.7/−2.7	−2.7/−3.7	−2.8
−9.0	1.0	−0.7	−2.7/−3.7	−3.7/−4.7	−3.8

Growth before manufacturing 0.1 log, growth at 5°C is 1.0 log/week, shelf life 3 weeks, MC = 0.1 PO, r-value in exponential dose-response model = −10, serving size = 50 g.

in fresh-cut lettuce (Table 2). The food is harvested and quickly processed. It has 1 week in commercial storage and 2 weeks of postretail storage before consumption. The growth rate at 5°C storage is 1.0 log per week. The MC parameters are such that the lot will be accepted (i.e., the PO is not exceeded) 95% of the time if the log(lot mean) is one log below the respective PO. In this example the serving size is assumed to be 50 g, and the exponential dose-response model estimates for the target population that consumption of 10,000 pathogens has a probability of causing one illness every million servings.

In this example, an LOP of one illness in one million servings (−6 $\log_{10}$) would have an FSO of 4.0 $\log_{10}$ CFU/serving. This corresponds to 2.3 $\log_{10}$ CFU/g at consumption. Allowing for 2 logs growth during postretail storage requires the PO at retail sale be 0.3 $\log_{10}$ CFU/g and the MC lot mean at the point would be −0.7 $\log_{10}$ CFU/g. The POs after manufacturing and for raw lettuce would need to be −0.7 and −0.8 $\log_{10}$ CFU/g, respectively. If another LOP were designated as the ALOP, the FSO, PO, and MC values would correspondingly change as shown in the table. As the ALOP becomes more stringent, the allowable contamination levels may become unobtainable. It is also important to note that there is a point below which verifying a PO with an MC becomes unfeasible due to the large number and/or size of the samples required by the sampling plan.

Appropriate level of protection. The level of protection deemed appropriate by the member (country) establishing a sanitary or phytosanitary measure to protect human, animal, or plant life or health within its territory (WTO, 1995). Other terms for the concept behind the ALOP have been acceptable level of protection and tolerable level of protection. This reflects the struggle to communicate that risks will never be zero. However, this does not mean that illnesses or deaths are "acceptable." There is also the realization that setting practical and feasible goals for industry and regulatory agencies will further food safety more than setting unrealistically and unobtainably low specifications. The ALOP is based on multiple factors including the pathogenicity of the microorganism, likelihood of the food being contaminated, feasible control measures, sensory and nutritional quality, costs, legal requirements, and consumer food preferences. In the end, the ALOP is a value judgment, not a scientific decision.

The objective articulation of an ALOP could potentially take many forms. It could be expressed in terms of an incidence of illnesses, hospitalizations, or deaths (i.e., number of cases per population per year). Alternatively, it could be expressed as the risk of illness per serving of food. The ALOP could also be based on the level of protection needed for specific populations at increased risk (e.g., children, elderly, immunocompromised consumers).

Food safety objective. The maximum frequency and/or concentration of a hazard in a food at the time of consumption that provides or contributes to the appropriate level of protection (Codex Committee on Food Hygiene [CCFH], 2005). The FSO can be considered a point value or bright line that no serving should exceed (ICMSF, 2002). It recognizes that the hazard characterization uses epidemiological and virulence data to create the dose-response model and supports decision making in a public arena on an agreed-upon ALOP. In contrast, the exposure assessment is a technical evaluation by a food processor or regulatory agency to determine the contamination levels for a specific food process. The FSO is the common point for these two different activities and areas of expertise. A bright line facilitates communication between public health agencies, industry, and consumers, and it provides an unambiguous regulatory criterion. The bright line focuses on the risk per serving and means that the high doses are always unacceptable, even though they may be very rare.

Performance objective. The maximum frequency and/or concentration of a hazard in a food at a specified step in the food chain before the time of consumption that provides or contributes to an FSO or ALOP, as applicable (CCFH, 2005).

Performance criterion. The effect in frequency and/or concentration of a hazard in a food that must be achieved by the application of one or more control measures to provide or contribute to a PO or an FSO (CCFH, 2005). This is the outcome of a process step or a combination of steps, i.e., the change in the level of a microorganism or microbial toxin (Stringer, 2005). The effect in frequency and/or concentration of a hazard in a food that must be achieved by the application of one or more control measures to provide or contribute to a PO or an FSO (CAC, 2004).

Microbiological criterion. The acceptability of a product or a food lot, based on the absence or presence, or number of microorganisms per unit(s) of mass, volume, area, or lot (CAC, 1997). The purpose of the microbiological criterion is to determine whether a lot of food is acceptable (pass) or not (fail). It is one of the means by which one verifies that a PO (or FSO) is being achieved to a specified level of confidence. The two-class attribute test is the most widely used type of MC. It is characterized by the number of samples to be tested (n), the number of samples (c) that exceed the criteria (in microbial testing associated with pathogenic microorganisms, c is typically zero), the lower limit of detection for the test (m), and a confidence level that an unacceptable lot is identified (e.g., 95%) (ICMSF, 2002; Dahms, 2004; Whiting et al., 2006).

Process criterion. A control parameter(s), e.g., time, temperature, pH, aw, at a step that can be applied to achieve a performance criterion (Stringer, 2005).

USING RISK ASSESSMENT TO ESTABLISH FOOD SAFETY
Risk Analysis
The cyclic process of identifying a problem, conducting an analysis, evaluating potential changes (mitigations), choosing and implementing a mitigation, and evaluating the impact of that mitigation is the risk analysis process (Dennis et al., 2008 [see chapter 5]). Three groups participate in the risk analysis: risk managers, risk assessors, and risk communicators (IOM/NRC, 2003). The overall risk analysis process is directed by the risk managers because they are responsible for making and implementing the decisions. The risk managers need to define an explicit question(s) and request particular outputs from the risk assessors. The risk assessors conduct the technical/scientific risk assessment. They collect and evaluate data, assemble process models, and calculate the results. At the direction of the risk managers, the risk assessors evaluate the impact of any mitigation that might be considered. Based on the results of the risk assessment and other regulatory, feasibility, economic, and value considerations, the risk managers decide on mitigations (if any), implement the changes, and monitor the impact of these changes. If the food process still does not meet food safety goals (FSO or ALOP), the risk managers may initiate another risk analysis cycle. Risk communication involves exchange of data and other information between risk managers, risk assessors, and various stakeholder and public groups. It is not limited to what

the risk managers are going to say to the consumer; it involves the entire communication matrix that is needed to design and implement a food safety initiative. Although it is crucial that the risk assessors remain free to produce an unbiased, scientifically sound risk assessment, experience has shown that frequent communication needs to occur between all groups to ensure that the appropriate data and processes are modeled, that the risk assessment provides answers that meet the needs of the risk managers, and that all groups understand the outputs and insights the risk assessment provides (Dennis et al., 2008 [see chapter 5]).

Thus, risk analysis provides the information and analysis that underlie the HACCP plan. After the decision that a process meets the ALOP, an appropriately designed risk assessment can help identify the CCPs and establish their critical control limits (HACCP steps 2 and 3). The risk assessment can also estimate the consequences of out-of-control processes and helps establish corrective actions. Without the risk analysis/assessment process, the HACCP system would have no metric for success or failure that is grounded in the level of risk that needs to be controlled. There would be no public health-based criteria underlying whether a process was "safe."

Risk Assessment

In the process of linking the entire processing–consumption–illness progression, risk assessors consider the relevancy and distribution of the data used in the assessment. Relevancy is an evaluation of how closely a data set represents the data needed for the risk assessment and risk management decision. For example, in modeling the risk of *L. monocytogenes* in deli meats (FDA/FSIS, 2003), contamination frequency and levels at retail were taken from an excellent survey conducted in 2000 in Maryland and northern California. A judgment must be made on the relevancy of those data to provide regulatory guidance in 2006. From 2000 to 2006, improved sanitation programs by the industry may have reduced contamination or an increased use of inhibitors such as sodium lactate and sodium diacetate may have reduced *L. monocytogenes* growth in the foods. Either change reduces the accuracy, thus relevancy, of the earlier data for current decision making. All survey data reflect the particular time and location of the survey. Experimental data are not as subject to becoming less relevant as survey data, but the risk assessor must always consider whether the strains used in an experiment represent all possible strains, how closely the laboratory or pilot plant reproduces the commercial processes, and whether, for example, the fixed elements in a temperature experiment (e.g., pH or salt level) are similar to the situation in the risk assessment to which the data are being applied.

A simple risk assessment or the FSO model above (Table 2) uses the average values for the model's parameters to make an estimate of the average rate of illness or expected number of illnesses. This is termed a deterministic model. However, most parameters exist as a distribution, i.e., oven temperatures are not uniform, contamination of raw ingredients is variable, and storage times are not the same for all packages from a production lot. *On average* our foods are "safe"; it is a small fraction of the servings that lead to illness. Alternatively, a deterministic risk assessment may use data values that represent the tails or worst case, e.g., select as the portion size for a food the amount consumed by the individual eating the largest amount. Depending on where on underlying distribution of values the model parameters have been selected, a deterministic risk assessment will have the potential of producing risk estimates that are insufficiently or inappropriately stringent. This is true in particular for a worst-case analysis that combines several worst-case values. Combining parameters in this manner rapidly produces an unlikely extreme estimate that is typically unrealistically stringent in terms of the true underlying risk. In recognition of this, risk assessments are increasingly using probabilistic modeling approaches. This allows the risk assessments to utilize the entire distribution of the data, not just the average. A probabilistic risk assessment, in effect, calculates the likelihood that combinations of values from each of the distributions occur to lead to a high likelihood of that serving causing illness.

The risk assessment uses the distributions of input values to calculate a distribution for the output (risk of illness). To do this, a repetitive calculation process, termed Monte Carlo analysis, is used where each repetition (iteration) takes one value from each of the distributions and calculates the expected number of pathogens per serving or risk of illness per serving. Because each repetition will take a different set of values, the outputs will create a distribution instead of a single value. Both the output mean and distribution are interpreted. The values at one or two standard deviations above the mean representing the 84th and 98th percentiles, respectively, may be more important for making a policy decision than the value for the mean. Sensitivity analyses can determine what parameters have the most impact on the outcome, both on the absolute value of the output distribution and on which parameters have high variation and/or uncertainty that contribute the most to the spread in the output distribution. The risk assessment will present a more complex answer to the risk manager's questions than traditional subjective or deterministic calculations but it will be a more complete answer.

The distribution of values within a data set originates from variability and uncertainty. Variability is the real differences that exist between different samples, e.g., different packages from a lot are not all stored for the same time before consumption. Additional data or better measurements will not minimize or eliminate this variability. Uncertainty, on the other hand, is from the errors in sampling, testing, and modeling. Replicate plate counts from a lot are not exactly the same; each estimates the actual mean value. Additional replicates or more accurate and precise tests would reduce the uncertainty. Most data sets have a combination of variability and uncertainty. For example, a telephone survey of home storage times will reveal that different consumers have kept a food for different lengths of time. However, there are limitations in the survey from the limited number of responses, nonrandomness of the responders (only those at home), and accuracy of the responder's answer that contribute uncertainty into the correctness of the resulting distribution in portraying the larger population that the survey intends to represent.

An advanced risk assessment format separates uncertainty from the variability. For example, the variation of a storage time distribution may be characterized by a minimum of 0.1 day, most frequent value of 3 days, and a maximum value of 10 days. However, the true value for the most frequent value may only be known to be between 2 and 5 days. A two-dimensional risk assessment separates variation from uncertainty by having a nested set of distributions describe each factor; the initial distribution is the variation, and each parameter characterizing that distribution has an uncertainty distribution. To perform and analyze a two-dimensional risk assessment, a value for each uncertainty distribution is selected according to that distribution (selected randomly or by Latin hypercube sampling). With this specific set of uncertainty values defining the variation distribution, the risk assessment is run and an output distribution is obtained. Then another set of uncertainty values is selected that define the variation distributions, and the risk assessment is rerun to obtain another output distribution. Figure 2 shows seven outputs; each sigmoidal curve is the cumulative plot of the output using one set of uncertainty values. The placement of the curve indicates the general risk level; the 5, 50, and 95% levels of risk can be readily viewed. The steepness of the curve reflects the variation; a relatively steep curve has low variation compared with a sloping curve. The spread between the different curves is an indication of the impact of the uncertainties. If the uncertainty is relatively small, the curves are close together; if they are widely spread relative to the slopes of the individual curves, then the uncertainty is high compared with the variation.

Another type of uncertainty is in the design of the process model. All models must be a simplification of the multiple pathways that a food may undergo, in particular, storage periods after manufacture. It may be unknown whether all the important steps have been put into the model. There may be a source of recontamination that is important but was not included in the

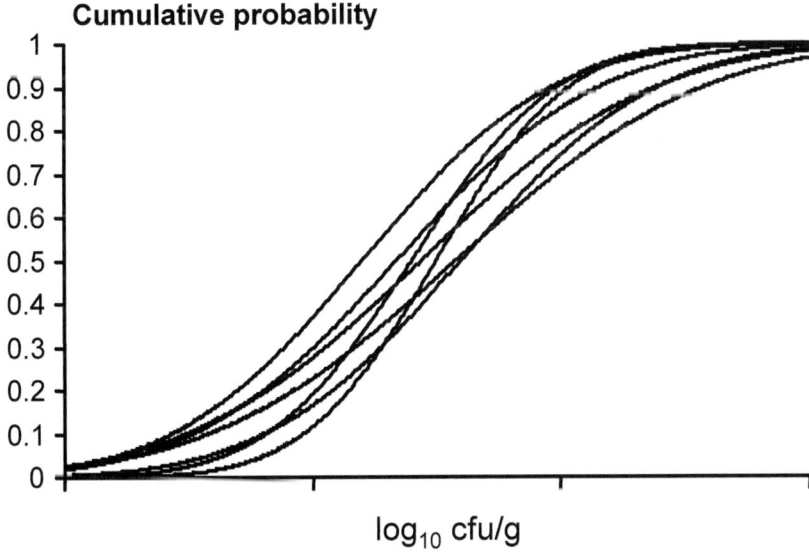

Figure 2 Illustration of the output of a two-dimensional risk assessment.

calculation, thereby making the model inaccurate by an uncertain amount. Individual growth or survival models for the different process steps also have uncertainties in their abilities to accurately represent the food-processing step, in particular, if broth-based data were used to create the model and the model has not been validated for that particular use.

Uncertainty also arises from the probabilistic nature of the sampling process. For example, with present/absent parameters, if a small number of samples were taken, the resulting percentages will have a relatively high uncertainty. A survey of 25 packages finds 3 contaminated packages; the 12% contamination rate (3 of 25) has a 95% confidence range from 4.4 to 30.2% for what the true value might be. However, if 100 samples had been taken and 12 packages were found contaminated, the narrower confidence range would be from 7.0 to 19.8% (beta distribution). Other distributions frequently encountered are the Poisson distribution for the number of microorganisms in a sample and a binomial distribution for the number of positive responses in a set of samples, given a known probability of being positive.

SETTING AN ALOP, FSO, PO

At the time of writing this chapter, no food regulatory agency has set a specific quantitative ALOP and, thereby, established a public process to do that (however, governments traditionally have made decisions about the safety of foods and implemented regulations to reduce the risk of illness). There are several approaches to setting an ALOP; one begins with the risk manager's determination that the rate of illness is unacceptably high for a food or class of foods and can be reduced. The new ALOP could be a judgment about how much lower the risk can feasibly be reduced at an acceptable cost. Feasibility involves the ecological, cultural, and virulence characteristics of the pathogen and what PO a well-designed and properly controlled food process can achieve. It considers the level of compliance with current and potential control measures. Costs to be considered include loss of sensory quality and nutritional value. Cultural food traditions, perceptions of risk, and legal and regulatory systems all affect the setting of the ALOP. A more severe thermal process, for example, may result in food that would have a very low risk but would be unacceptable to consumers. The public may demand continued access to higher risk foods, e.g., sunny-side-up fried eggs, unpasteurized goat milk, raw oysters, and cheeses made from unpasteurized milk. The public may or may not accept process changes that reduce the risk depending on added costs and sensory changes. The commercial acceptance of irradiated foods and pasteurized shell eggs is currently unsure despite their acknowledged ability to reduce the risk of food-borne illness. Whether the consumer should be allowed a choice or whether the regulatory agency should mandate a safety-increasing process is determined by the cultural and legal traditions of a country.

The decision about an ALOP, of necessity, often focuses on the susceptible human populations. Children are susceptible to *E. coli* O157:H7 and pregnant women and their fetuses are susceptible to *L. monocytogenes*. Consumption of foods cannot be easily restricted from certain groups, and the ALOP will need to reflect the more susceptible populations. However, there are some extremely susceptible populations, such as organ transplant and bone marrow transplant patients, who would require an LOP that may not be feasible for some foods. These groups may have to be protected by other measures. The severity of the illness also affects the degree of measures that would be accepted to avoid the illness.

Economic analyses have not, thus far, been a formal part of the risk analysis paradigm. However, cost-benefit analyses of proposed regulations are part of the U.S. and other countries' regulatory systems. Evaluating costs and safety with the goal of achieving the greatest degree of safety for a given expenditure has been explored (Williams and Thompson, 2004). If an economic analysis will be done, the risk assessment should be designed from the beginning to provide appropriate outputs for the economists.

Using Risk Assessment To Analyze and Improve a Food Process

A risk assessment for a food process would likely be initially approached with a deterministic model, then one-dimensional risk assessment, and finally a two-dimensional risk assessment (van Gerwen et al., 2000). This provides the greatest assurance of appropriate modeling and data usage. With the development of a valid two-dimensional process model, the model can assist in improving a food process or determining whether it meets an FSO.

The general order for improving the process would be to first reduce the uncertainties to determine whether the process was actually acceptable. Figure 3a shows the output of a process risk assessment showing three of the uncertainty runs for a process having high variation and uncertainty. If the vertical dashed line at 1.8 $\log_{10}$ CFU/g represents the PO that would meet the FSO, this process would be judged to be unacceptable because a significant fraction appears to exceed the PO. However, the process may, in fact, meet the PO if the uncertainty values for the steepest variation curves represent the "true" distributions. In this case, no modifications of the process would be necessary. However, if the other variation curves represent the "true" distribution, then the process is clearly unacceptable. Sensitivity analyses can identify the specific parameters whose uncertainty makes the greatest contribution to the overall uncertainty. Reducing uncertainty requires collection of additional data to refine the distributions; for example, temperature readings are made throughout the oven to precisely measure cold spots. With the new information, the risk assessment is recalculated and Fig. 3b might result. This shows that most of product is acceptable but a sufficiently high probability remains that some of the servings exceed the PO. If the process is considered unacceptable, then a sensitivity analysis for variation identifies the parameter(s) that significantly contributes to the output distribution. If an oven has hot and cold spots, improving the air circulation within the oven would reduce the distribution in temperatures and resulting thermal inactivation. Better inventory control could reduce the distribution of storage times. Reducing variation would likely improve the consistency of the sensory and nutritional qualities of the product as well as the microbial consistency. Reducing the variation in the process steps that make the largest contribution to variation, collecting new data, and redoing the risk assessment may result in Fig. 3c. If the process still has an unacceptable portion of trials (servings) that exceed the PO, the process must be changed (performance criteria or process criteria) to lower the absolute value of the output distribution (Fig. 3d). For the oven, the temperature could be increased so that all of the individual products receive a greater thermal inactivation.

This is the general order of assessing and correcting a food process; however, the evaluation of individual processes may differ. For example, it may be immediately recognized that reducing the uncertainty of consumer practices will be difficult or that a food can easily be given additional thermal

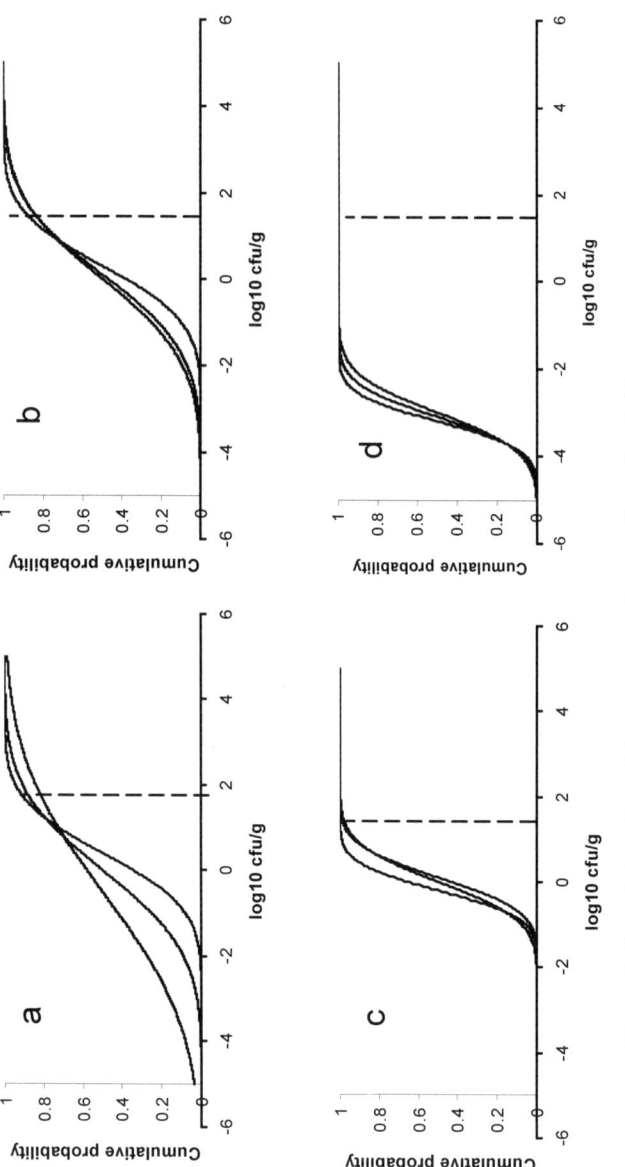

Figure 3 Cumulative plots of the outcomes of a two-dimensional risk assessment.

processing without significant sensory degradation or additional cost. Both uncertainty and variation could be analyzed at the same time rather than sequentially.

Validation

The iterative risk analysis procedure described above would lead to the design of a process that is capable of meeting the FSO/PO. Because the data for the risk assessment most likely came from a variety of sources, including published laboratory studies and pilot plant trials, the process needs to be validated to ensure that the FSO will be met in the actual food-processing operation. This is not always possible to demonstrate directly as pathogens cannot be brought into the processing plant and many of the performance objectives may be too low to be measurable. Validation may need to be partial and use surrogate microorganisms. The analysis of sufficient product lots should be taken to establish that the normal operating characteristics of the process achieve the intended level of control.

Verification

Verification means ensuring that the process is operated according to the design parameters (HACCP plan). Redundancy in monitoring and control of critical control points can reduce out-of-process events. Avoiding human mistakes through job structuring, training, and proper motivation is probably the key to avoiding failure of a properly designed food-processing operation. As with the initial validation of the process, appropriate samples need to be taken and compared with the established performance expectations (Shewart plots) to verify that the process remains in control (Wheeler and Chambers, 1992).

SUMMARY

Traditional assessments of the microbial hazards in foods were based on data for the characteristics of the pathogen, contamination levels, effects of process steps, ability of the pathogen to grow in the food, and the epidemiological record. These data were treated as independent data sets and subjectively interpreted by policymakers. Quantitative microbial risk assessment extends the mathematical treatment of the data and food process to include all of the relevant process steps. Growth, survival, or inactivation models are applied to each step and the steps are linked from raw material to the consumer. A dose-response model then relates consumption to the probability of illness or other measure of public health. Thus, the risk assessment links the process parameter values (product pH, storage temperature) to public health (illnesses per serving).

This process needs information for every processing step and requires risk assessors to evaluate the quality and relevancy of the data available to them. A data set has individual measurements, a mean, a distribution of the measurements (variation), and an estimate of how precise the mean and other descriptors of the distribution are to the population values needed for the risk assessment (uncertainty). Analyses of the distributions for the process parameters and output in public health permit interpretations about the frequency of specific outcomes. The frequency that the risk exceeds a specific rate (e.g., 2% of the servings exceed a risk of one case per million servings) can be more valuable information for risk managers than the average risk per serving.

Estimating and separating the contributions of the process performance, variation in process parameters, and knowledge uncertainty can provide insights for guiding approaches to remediation. In general, if a process appears to be unacceptable but the data are poor (high uncertainty), the initial effort should be to collect better data and reanalyze the process. If the process is still unacceptable, then ways to reduce the variation should be sought and, after collecting new data, the process is again reanalyzed. If the process is still unacceptable, then the process may need to be redesigned or the performance objective for one or more steps changed to reduce the risk. Thus, a risk assessment helps determine the parameter values and their acceptable ranges for a process to meet a food safety target. These values become the critical control points in the HACCP system that plant personnel use to monitor and control the food process.

REFERENCES

Buchanan, R. L., and R. C. Whiting. 1998. Risk assessment: a means for linking HACCP plans and public health. *J. Food Prot.* 61:1531–1534.

Codex Alimentarius Commission (CAC). 1997. Principles for the establishment and application of microbiological criteria for foods, CAC/GL 21-1997. Secretariat of the Joint FAO/WHO Food Standards Programme, Food and Agriculture Organization of the United Nations, Rome, Italy.

Codex Alimentarius Commission (CAC). 2004. ALINORM 04/27/13, Appendix III, 9p. 83. Food and Agriculture Organization of the United Nations, Rome, Italy.

Codex Committee on Food Hygiene (CCFH). 2005. Proposed draft principles and guidelines for the conduct of microbiological risk management (MRM). ALINORM 05/28/13, Appendix III.

Dahms, S. 2004. Microbiological sampling plans—statistical aspects. *Mitt. Lebensm. Hyg.* 95:32–44.

Dennis, S. B., J. Kause, M. Losifkoff, D. L. Engeljohn, and R. L. Buchanan. 2008. Using risk analysis for microbial food safety regulatory decision making, p. 137–175. *In* D. Schaffner (ed.), *Microbial Risk Analysis of Foods*. ASM Press, Washington, DC.

Food and Agriculture Organization/World Health Organization (FAO/WHO). 2002. Principles and guidelines for incorporating microbiological risk assessment in the development of food safety standards, guidelines and related texts. Report of a Joint FAO/WHO Consultation, Kiel, Germany. Food and Agriculture Organization of the United Nations, Rome, Italy.

Food and Agriculture Organization/World Health Organization (FAO/WHO). 2003. *Hazard Characterization for Pathogens in Food and Water*. Microbiological Risk Assessment Series no. 3. Food and Agriculture Organization of the United Nations, Rome, Italy.

Food and Drug Administration/Food Safety and Inspection Service (FDA/FSIS). 2003. *Quantitative Assessment of Relative Risk to Public Health from Foodborne* Listeria monocytogenes *among Selected Categories of Ready-to-Eat Foods*. www.foodsafety.gov/~dms/lmr2-toc.html.

Holcomb, D. L., M. A. Smith, G. O. Ware, Y.-C. Hung, R. E. Brackett, and M. P. Doyle. 1999. Comparison of six dose-response models for use with food-borne pathogens. *Risk Anal.* 19:1091–1100.

Institute of Medicine/National Research Council (IOM/NRC). 2003. *Scientific Criteria to Ensure Safe Food*. The National Academies Press, Washington, DC.

International Commission on Microbiological Specifications for Foods (ICMSF). 2002. *Microorganisms in Foods*, vol. 7. *Microbiological Testing in Food Safety Management*. Kluwer Academic/Plenum Publishers, New York, NY.

McKellar, R. C., and X. Lu. 2004. *Modeling Microbial Responses in Food*. CRC Press, Boca Raton, FL.

Nauta, M. J. 2008. The modular process risk model (MPRM): a structured approach to food chain exposure assessment, p. 99–135. *In* D. Schaffner (ed.), *Microbial Risk Analysis of Foods*. ASM Press, Washington, DC.

Ross, T. 2008. Microbial ecology in food safety risk assessment, p. 51–98. *In* D. Schaffner (ed.), *Microbial Risk Analysis of Foods*. ASM Press, Washington, DC.

Stringer, M. 2005. Food safety objectives—role in microbiological food safety management. *Food Control* 16:775–794.

van Gerwen, S. J. C., M. C. te Giffel, K. van 't Riet, R. R. Beumer, and M. H. Zwietering. 2000. Stepwise quantitative risk assessment as a tool for characterization of microbiological food safety. *J. Appl. Microbiol.* 88:938–951.

Wheeler, D. J., and D. S. Chambers. 1992. *Understanding Statistical Process Control*, 2nd ed. SPC Press, Inc., Knoxville, TN.

Whiting, R. C., and R. L. Buchanan. 2001. Predictive modeling and risk assessment, p. 728–739. *In* M. P. Doyle, L. R. Beuchat, and T. J. Montville (ed.), *Food Microbiology: Fundamentals and Frontiers*, 2nd ed. ASM Press, Washington, DC.

Whiting, R. C., and R. L. Buchanan. 2007. Progress in microbiological modeling and risk assessment, p. 953–969. *In* M. P. Doyle and L. R. Beuchat (ed.), *Food Microbiology: Fundamentals and Frontiers*, 3rd ed. ASM Press, Washington, DC.

Whiting, R. C., A. Rainosek, R. L. Buchanan, M. Miliotis, D. LaBarre, W. Long, A. Ruple, and S. Schaub. 2006. Determining the microbiological criteria for lot rejection from the performance objective or food safety objective. *Int. J. Food Microbiol.* **110**:263–267.

Williams, R. A., and K. M. Thompson. 2004. Integrated analysis: combining risk and economic assessments while preserving the separation of powers. *Risk Anal.* **24**:1613–1623.

World Trade Organization (WTO). 1995. The WTO agreement on the application of sanitary and phytosanitary measures (SPS Agreement). Available at www.wto.org/English/tratop_e/sps_e/spsagr_e.htm.

Microbial Risk Analysis of Foods
Edited by Donald W. Schaffner
© 2008 ASM Press, Washington, D.C.

Microbial Ecology in Food Safety Risk Assessment

3

Tom Ross

INTRODUCTION

The health risks associated with food- and water-borne microbiological hazards are influenced by a complex interplay of variable factors. In this chapter, risk is assumed to relate directly to risk to human health, considering the probability and severity of illness both to individuals and the overall population exposed. The virulence of the pathogen (i.e., the set of characteristics that enable it to cause disease), the susceptibility of the consumer to those virulence determinants, the probability that the pathogen is present in the food at all, and the number of cells of the pathogen that are ingested (a function of serving sizes, the extent of contamination, and the handling and formulation of the food) will all affect the probability and severity of illness associated with that pathogen in that food. While the risk to a population is dictated by frequency of contamination and the distribution of doses, the probability of infection of an individual consumer is ultimately based on the number of pathogens ingested.

In principle, frequencies of contamination and numbers of cells ingested could be measured at, or near, the point of consumption, but this is highly impractical and such data are usually not available. Instead, data relating to pathogen levels and frequencies at some earlier stage in the food chain are used to *infer* frequencies and levels that could be expected to be present at the time that the food is eaten. This inference is based on the conditions that the food, and the pathogens within it, experience between the point in the food chain to which the data relate and the point at which the food is consumed, often using the techniques and tools of "predictive microbiology." This

Tom Ross, Australian Food Safety Centre of Excellence, School of Agricultural Science, Private Bag 54, University of Tasmania, Hobart, Tasmania 7001, Australia.

estimation process will be an essential part of the exposure assessment in many microbial food safety risk assessments where (i) microbiological data relevant to the food at point of consumption are not available or (ii) the purpose of the risk assessment is to identify the best strategies and actions to be applied at specific processing or handling steps so as to reduce the risks associated with consumption of the food.

Vose (1996) offered as his cardinal rule of risk analysis employing stochastic simulation modeling that: "Every iteration of a risk analysis model must be a scenario that could physically occur." For the microbial risk assessor, this dictum presents a strong challenge because microorganisms, with the exception of viruses, are living things. As such, compared with other kinds of food-borne hazards (e.g., chemical residues, biological toxins, allergens, foreign objects), they can be exquisitely sensitive to the environmental conditions that they encounter in foods. Indeed, the food industry and consumers go to great lengths to try to impose environmental conditions in foods and during food processing and distribution that either:

- prevent pathogens from getting into or onto foods,
- remove them from foods (e.g., filtration),
- inactivate them, if present, in foods (e.g., with heat or chemicals) or, at the very least,
- minimize their growth, if present (e.g., with chemical preservatives, by refrigeration).

The effect of microorganisms is related more to their number than their size, and microbiologists typically think in terms of populations of microbes. Thus, when microbiologists talk about microbial growth they are usually referring to the increase in the size of a population. The term "proliferation" may be more appropriate and is used in this chapter. The study of the effects of environmental conditions on microorganisms is termed "microbial ecology." From the perspective of food-borne microorganisms in a unit of food, that food is an environment that can present opportunities for growth, or pose a threat to survival, depending on the physicochemical conditions in the food and the conditions of storage. From the preceding discussion, it might be expected that development of reliable microbial food safety risk assessments will require a sound understanding of the microbial ecology of foods and the ability to express that understanding mathematically.

As noted, microbial hazards in foods can be extremely sensitive to their surroundings and, under some conditions, the microbial ecology of foods has characteristics of a chaotic system. However, while the process of meaningful risk assessment of microbial pathogens in foods is not trivial (in particular, the modeling of the dynamics of pathogen populations along the farm-to-fork chain), the behavior of microbial pathogens in foods is certainly not a "black box." The microbial ecology of foods is deterministic and much *is* known that can be applied to microbial food safety risk assessment to increase the scientific credibility and utility of the risk assessment outcomes. Traditionally, it was considered that microbial responses in foods under temporally changing conditions were highly unpredictable and that food microbiology was "more art than science." In recent decades, however, food microbiologists have adopted an increasingly quantitative and mechanistic approach to description of the microbial ecology of foods, this approach first having been advocated and termed "predictive microbiology" by Roberts and Jarvis (1983). The ability to predict microbial behavior in foods is suggested to have been a critical element in the extension of the risk assessment approach to microbial food safety, a process initiated in the early to mid-1990s.

Equally, however, as microbiologists have delved more deeply into the mechanisms underlying the responses of populations of microorganisms in foods, have addressed more complex systems, and have begun to quantify the variability in responses between strains of single pathogenic species, it has become apparent that there *are* circumstances where the fate of the microorganism in a food *may be* inherently unpredictable.

One of the reasons for this unpredictability is the capacity of microorganisms for self-amplification in foods or in consumers in the process of infection. Thus, a single cell in a 100-g serving of food—representing a contamination level of 1 in 10^{14} (w/w)—has the capacity to grow and make the entire serving unsafe for consumption. Equally, microorganisms exist as discrete particles of size much larger than chemical toxins and, unless initially present at high levels, will not be partitioned homogeneously between individual units of foods. Thus, some units of food drawn from a single batch may develop disease-causing levels of microbes or microbial toxins while the majority of the batch remains completely safe. Compounding these difficulties are variability in microbial responses due to slight and random differences in their genetic potential, the inherent uncertainty concerning which organisms are initially present in the unit of food, and the conditions they experience in the food both in time and in space prior to consumption.

This chapter aims to (i) summarize knowledge of the deterministic aspects of microbial ecology of foods that is relevant to the conduct of microbial food safety risk assessment, (ii) highlight sources of variability and uncertainty in microbial behavior, (iii) review approaches that have been adopted to model microbial growth, stasis, and death along the farm-to-fork pathway, and (iv) identify sources of information that can assist microbial food safety risk assessors to produce models that are more scientifically rigorous and thus defensible. Sections of this chapter draw on a recent review (Ross and McMeekin, 2003) but have been updated. Additional topics, including attention to elements of the microbial ecology of foods that are variable or uncertain, are also introduced and discussed.

OVERVIEW OF MICROBIAL PATHOGENS

Five main groups of microorganisms are recognized:

- algae
- fungi
- protozoans
- bacteria, and
- viruses.

Each group contains pathogens that may be present in foods and lead to infections in humans or that may produce toxic compounds that can persist in foods. One of the unique features of microbial risk assessment, in comparison with assessment of the risk from chemical hazards, is that microbial hazards all have the capacity to "self-amplify," that is, to produce more copies of themselves or increased levels of toxic compounds.

Prions are a more recently described class of disease agent that have the characteristics of an infectious agent. Prions, however, differ from the groups above because they do not contain genes composed of ribonucleic acids and, as such, are not an "organism." They are associated with a range of diseases of mammals collectively termed "transmissible spongiform encephalopathies" (TSEs), which include "mad cow disease" or bovine spongiform encephalopathy (BSE), kuru, Creutzfeld–Jakob disease (CJD), variant Creutzfeld–Jakob (vCJD), scrapie, and others. The prion is hypothesized to be a protein molecule normally associated with mammalian cell membranes that has adopted a conformation (three-dimensional structure) that makes it unable to fulfill its correct biological function. A feature of the prion is that it is able, by a mechanism not yet fully known, to catalyze the conversion of the normal protein to the prion form. As such, the prion does share the property of self-amplification, characteristic of infectious agents, but only when in contact with the normal protein, which is, in effect, the substrate for production of more prion.

The diversity of physiology and ecology among the five groups of microscopic organisms is far greater than the diversity among all plant and animal species. Indeed, the genetic diversity among just one of those five groups (the *Bacteria*) is greater than that of all the known plants and animals combined. Fortunately, the number of organisms among these groups that are recognized as food-borne hazards is relatively small, and there are principles of their ecology that are common to all groups. In this section an overview of characteristics unique to each group of organisms is given, focusing on the way they behave in foods and the type of food safety hazard they present.

Algae

Relatively few algae are associated with food-borne illness, but up to 80 species of single-celled microalgae produce toxins with effects on humans ranging from gastrointestinal illness, through to paralysis, severe neurological effects, and even death. Examples include *Pseudo-nitzschia* spp., *Gambierdiscus toxicus*, *Prorocentrum* spp., *Ostreopsis* spp., *Coolia monotis*, *Thecadinium* spp., *Amphidinium carterae*, *Dinophysis* spp., *Karenia brevis*, *Alexandrium* spp., *Gymnodinium catenatum*, and *Pyrodinium bahamense*. In humans, most of the toxins exert their effect through disruption of sodium-ion channels in neurons required for transmission of nerve impulses.

The microalgae can be ingested and concentrated by filter feeding shellfish. The microalgae do not grow in the shellfish but contain toxins that are absorbed by the shellfish as they ingest the microalgae. It is the toxin in the flesh and organs of the shellfish that presents a human health hazard. The toxin does not further increase after the oyster has ingested the algae but is slowly eliminated from the live oyster through metabolism by the oyster (and possibly by resident bacteria). Typically, high numbers of cells ($<10^4 \, g^{-1}$) in shellfish-growing waters are required to lead to accumulation of toxin sufficient to cause overt disease in human consumers. The correlation between algal levels in the water and toxin levels in shellfish is poor; but for some species of toxic microalgae, the hazards are considered to be so great that their mere presence in the shellfish-growing area is enough to require those shellfish to be tested for the presence of toxin: levels in the shellfish are used to determine acceptability for harvest or sale. The toxins are all tasteless and odorless and are very stable to acid and heat, so that they are not reliably eliminated by cooking.

One of the most common reported seafood-borne illnesses in humans is ciguatera poisoning. Ciguatera toxin is derived from dinoflagellate microalgae, with *G. toxicus* being the most studied and most frequently implicated. It is an epiphytic (i.e., nonplanktonic) organism that lives on surfaces of coral and seaweeds on tropical reefs and is ingested by grazing fish. The toxins contaminate, and persist in, the flesh of the fish. Those fish will be ingested by larger fish, and so on, so that the toxin in passed to higher trophic levels and becomes concentrated in the larger (and older) predatory reef fish. Such fish are prized among amateur fishers. Eating these contaminated fish causes a variety of symptoms in humans, mostly mild, but sometimes extending to long-term neurological disturbances. The toxin is stable once ingested by the grazing fish, so that the microbial ecology of *G. toxicus* is unlikely to be a major part of a "boat-to-throat" risk assessment for ciguatoxin poisoning. Nonetheless, there is some understanding of the ecological factors that increase the likelihood of *G. toxicus* being present at a particular reef. This knowledge could potentially be exploited to identify fishing areas presenting greater probability of ciguatoxic fish as part of a management strategy involving education of fishers.

Fungi

The fungi include yeasts and molds. Yeasts are capable of anaerobic growth and normally exist as free-living single cells. Molds require oxygen for growth and tend to exist as multicellular aggregations, characterized by the formation of long chains of cells, called hyphae. When sufficient cells are present to be seen with the unaided eye, mold colonies usually appear "furry."

Yeasts are not usually associated with food-borne illness. Molds, however, may produce a range of toxins as they grow (i.e., proliferate) in, or on, foods. Fungi and bacteria are the only groups of microorganisms able to grow in foods. Fungi grow more slowly than bacteria and, as a result, fungal contaminants usually present problems only in those foods in which they have an ecological advantage over bacteria. Molds as a group are more tolerant of dryness and acidity than bacteria and, for this reason, are of most concern in foods that have relatively low water activity (a_w) or relatively low pH. (Water activity is defined as the vapor pressure of a solution [or other material containing water] divided by the vapor pressure of pure water at the same temperature. Because of interactions between molecules, as solute concentration increases, the water molecules require more kinetic energy to escape the water phase to become vapor. Thus, water activity is a measure of the relative availability of water in a substance to participate in chemical reactions, including microbial metabolism.) In current food and agricultural practice, grains and nuts are the commodities most affected by the presence of fungal toxins. Dried fish products are also reported to harbor mycotoxigenic fungi and, potentially, their toxic metabolites.

"Mycotoxins" is the collective term for the toxins produced by molds. Mycotoxins include the groups aflatoxins, ochratoxins, patulin, fumonisins, and tricothecenes. These toxins produce a range of effects in the mammalian host with the primary sites of damage being the liver and kidney, but they also cause neurological effects, necrosis, or immunosuppression. Several of the groups of toxins are carcinogenic, with Aflatoxin B1 being recognized as the most potent carcinogen, and associated with liver cancer. Ochratoxin A causes kidney damage. The trichothecenes are acutely toxic to humans, causing sickness and diarrhea and, potentially, death. Members of the genera *Aspergillus*, *Fusarium*, *Penicillium*, *Stachybotrys*, and *Mycothecium* are the most important producers of mycotoxins. Toxin production is species and strain specific, and also affected by environmental conditions, requiring that a case-by-case approach be applied to risk assessment of specific toxins produced by molds or that the variability between strains be considered.

To date, risk assessment of mycotoxins has adopted the techniques of chemical risk assessment and has relied on survey data to quantify concentrations and frequencies of mycotoxins in foods, coupled with food consumption data, to derive estimates of human exposure. In general, mycotoxins are quite stable and, once formed in food, toxin levels are unlikely to be changed due to food processing other than by dilution, e.g., due to mixing with other components, or concentration, e.g., due to removal of water or separation of oils (note that the solubility of aflatoxins in water is low).

This approach has arisen, presumably, because mycotoxin contamination of foods often arises prior to harvest, a section of the food chain traditionally considered to be beyond the control of the "food" industry. The application of formal risk assessment approaches to microbial food safety management, however, has introduced a holistic approach including consideration of food safety interventions that can be applied "in the field." Equally, mycotoxin production occurs during storage of grains and seeds. Thus, for some risk assessments it will be important to be able to estimate factors affecting mycotoxin production. The relatively large number of factors known to affect myoctoxin production, as well as the abundance of possible combinations, complicates the planning of a predictive model for mycotoxin production in storage (Shapira, 2004), and very few attempts have been made to incorporate the dynamics of mold growth and mycotoxin production in risk assessments. Factors of relevance include the availability of specific nutrients and gaseous atmosphere. The most important factors, however, are water activity and temperature. In general, the temperature, water activity, and pH limits for mycotoxin production are somewhat narrower than the corresponding limits for growth of the mold producing the mycotoxin. Environmental limits to mold growth and mycotoxin production were reviewed by Sanchis (2004).

Protozoans and Helminths
Protozoans are single-celled animals. Helminths are (multicellular) worms and include several species that that can infect mammals. Collectively they are called parasites. Both groups contain species that can be transmitted via food, in particular, meat and fish. Many are passed from the host through the feces and, thus, they can contaminate water. Accordingly, many types of fruits and vegetables can become vehicles for transmission of parasitic infection through

irrigation, or washing during processing, with contaminated water. Within the helminths three main groups able to cause human illness are recognized:

- nematodes (roundworms)
- trematodes (flatworms or "flukes"), and
- cestodes (tapeworms).

Food-borne parasites do not grow in foods and, while they are susceptible to many food processing and preparation steps, if such processes are not adequately applied or not expected to be necessary, parasites can survive and remain infective in some foods until the point of consumption. Protozoa can produce resistant resting stages, termed cysts or oocysts, that are quite stable, but inert, in the environment. They can then develop in appropriate hosts, but, unlike the viruses (see below), the host-species range is often broader and can include both human and nonhuman species.

Helminths may be transmitted by foods as adult or larval worms, or as eggs. The infectious dose for parasites is often only a single organism, egg, or cyst. Thus, the ecology of these pathogens in foods is of little relevance, except when the conditions lead to inactivation (i.e., death) of the parasite. Such conditions include freezing, cooking, and chemical preservation due to pH or water activity reduction (e.g., pickling), but for these processes to be reliably effective they must be severe enough and applied for long enough. Data indicating the required severity and time are available in the published literature and are included in guidelines for control of these pathogens in foods known to be likely to have parasites present. Such foods, typically, are based on raw meat or fish such as pickled, marinated, or cold smoked fish, uncooked fermented meats, and so on.

An overview of food-borne pathogenic protozoans and helminths, and the diseases they cause, is given in Table 1.

Viruses

Viruses are obligate intracellular parasites and are usually highly species specific in the range of organisms that they can infect. To reproduce they must first find the appropriate host, invade host cells, and usurp the host cell's metabolism to create more copies of themselves. To undertake further reproduction they must then "break out" of the host cell and invade adjacent cells, disperse within the host body to invade other tissues, or be carried outside of the original host to infect a new host. Viruses do not have cell structure and, under some definitions of life, are not living.

Viruses essentially are parcels of genes encased in a protein coat. The coat enables them to survive outside of their preferred host but is also very important in recognizing and invading the host cell. Viruses do not grow in

Microbial Ecology in Food Safety Risk Assessment   59

Table 1 Food-borne parasites

Illness	Symptoms	Causative agent	Type	Foods associated
Ascariasis	Often mild; intestinal discomfort and blockage	*Ascaris lumbricoides; Trichuris trichiura*	Helminth	Eggs can reside in soil/water and be transmitted via vegetables eaten raw
Toxoplasmosis	Healthy people: mild "flulike" illness Immunocompromised: eye, brain infections; fetuses may be infected, leading to mental retardation or physical abnormalities	*Toxoplasma gondii*	Protozoan	Raw/undercooked meat
Trichinosis	Nausea, diarrhea, vomiting, fever, and abdominal pain, followed by headaches, eye swelling, aching joints and muscles, weakness, and itchy skin	*Trichinella spiralis*	Helminth	Raw/undercooked meat
Taeniasis	Usually without symptoms but abdominal pain, weight loss, digestive disturbances, and possible intestinal obstruction; perianal irritation	*Taenia saginata* (beef tapeworm) *Taenia solium* (pork tapeworm)	Helminth	Raw/undercooked beef Raw/undercooked pork
Anisakiasis	Tingling or tickling sensation in the throat; in more severe cases, acute abdominal pain with nausea	*Anisakis simplex* (herring worm); *Pseudoterranova* (a.k.a. *Phocanema, Terranova*) *decipiens* (cod or seal worm); *Contracaecum* spp.; *Hysterothylacium* (*Thynnascaris*) spp.	Helminth	Raw, undercooked, or lightly preserved fish (e.g., sushi, ceviche, cold smoked salmon, pickled herring)
Giardiasis	Diarrhea, cramps, nausea	*Giardia duodenalis* (form. *lamblia*)	Protozoan	Water, fruits/vegetables
Cryptosporidiosis	Watery diarrhea, stomach cramps, upset stomach, and slight fever	*Cryptosporidium parvum*	Protozoan	Water, fruits/vegetables
Cyclosporiasis	Watery diarrhea (sometimes explosive), stomach cramps, nausea, vomiting, muscle aches, low-grade fever, and fatigue	*Cyclospora cayetanensis*	Protozoan	Water, fruits/vegetables

foods and, as with the protozoans, helminths, and algae, the ecology of these pathogens in foods is of little relevance in the context of food safety risk assessment, except when the conditions lead to their inactivation (i.e., because their inactivation is expected to lead to a reduction in risk).

Food-borne viral pathogens can be classified into three main groups according to the type of illness they cause. First, there are the viruses that cause gastroenteritis, such as norovirus (previously termed Norwalk Like Viruses, abbreviated to NLVs) and rotavirus. The second group comprises enterically transmitted hepatitis viruses, specifically hepatitis A and hepatitis E. The third group consists of the enteroviruses, such as poliovirus and echoviruses. Enteroviruses can replicate in the mucosal cells of the human intestine but cause illness after they migrate to other organs, such as the central nervous system or liver, and are associated with a diverse range of clinical symptoms.

Enteric viruses are the most common pathogens transmitted via food, with recent estimates indicating that they account for 50 to 70% of food-related illnesses in developed countries. On the basis of the number of outbreaks and people affected, norovirus and hepatitis A virus are considered by many as the most important human food-borne pathogens in developed nations.

Bacteria

Bacteria are unicellular organisms. They are one of two forms of life classified as prokaryotes, the other being the *Archaea*. The cell structure of prokaryotes is very different from that of higher organisms, including algae, parasites, and fungi, which are eukaryotes. Prokaryotes are the smallest of life forms, typically being only a few microns long and a micron, or less, in diameter. In terms of their genetic diversity, the range and number of habitats they occupy, and on the basis of their total biomass on the earth, the *Bacteria* are the most abundant of all organisms. Because of their small size, simplicity of form, and relative hardiness, they are easily dispersed in air or water droplets and tend to contaminate many anthropogenic and natural environments. They colonize (i.e., proliferate in) any environment that provides their (often simple) requirements for growth unless steps are actively taken to eliminate them. For the same reasons that food supports human growth and metabolism, many foods support bacterial growth and much effort is expended to preserve and protect foods against colonization and spoilage by bacteria. As well as spoilage organisms, the *Bacteria* include a relatively small number of pathogens that cause food-borne disease (see Table 2).

There are three main modes by which bacteria can cause food-borne human illness:

Table 2 Food-borne bacterial pathogens

Causative agent	Type	Symptoms	Foods associated	Special characteristics
Aeromonas spp.	Infection	Gastroenteritis (septicemia in immunocompromised)	Fish, shellfish	Growth at refrigeration temperatures
Bacillus cereus	Intoxication Toxicoinfection	Nausea, vomiting Diarrhea	Starchy foods, rice, custards	Spore former
Brucella	Infection	Sweating, headache, fatigue, fever	Raw animal products (meat, shellfish, chicken, milk)	
Campylobacter jejuni	Infection	Diarrhea (occult blood), fever, headache	Undercooked chicken	Narrow temperature range for growth, microaerophilic
Clostridium botulinum	Intoxication	Neurological (paralysis)	Canned or bottled vegetables, vacuum-packed, lightly preserved fish, some indigenous meals	Spore former, strict anaerobe
Clostridium perfringens	Toxicoinfection	Abdominal cramps, diarrhea	Meats, meat products, and gravy	Spore former, strict anaerobe
Enterobacter sakazakii	Infection, invasive	Septicemia (among neonates, very immunocompromised)	Reconstituted milk powders	
Escherichia coli (various)	Infection	Gastroenteritis (bloody diarrhea); young may develop hemolytic-uremic syndrome (HUS), characterized by renal failure, elderly may develop fever and neurological symptoms in addition to severe gastroenteritis	Undercooked meat, hamburgers; salads cross-contaminated with raw meat or fecal-contaminated water	
Listeria monocytogenes	Infection	Septicemia, meningitis (or meningoencephalitis), encephalitis, intrauterine or cervical infections in pregnant women, which may result in spontaneous abortion (2nd/3rd trimester) or stillbirth	Long shelf life, perishable refrigerated foods, eaten without further cooking (processed meats, salads, fresh cheeses, etc.)	Growth at refrigeration temperatures
Salmonella	Infection	Nausea, vomiting, diarrhea, fever	Raw meats, poultry, eggs, coconut, peanut butter, chocolate	

(*continued*)

Table 2 *(continued)*

Causative agent	Type	Symptoms	Foods associated	Special characteristics
Shigella flexneri	Infection	Abdominal pain; diarrhea; fever; vomiting; blood, pus, or mucus in feces	Salads and raw vegetables through contaminated water, milk and dairy products, and poultry	
Staphylococcus aureus	Intoxication	Nausea, vomiting	Many; often cooked products that are subsequently rehandled and thermally abused	Unusually tolerant to salt/low water activity
Streptococcus pyogenes	Infection	Sore throat, scarlet fever, other septicemia, fever, tonsillitis, skin rash	Various, but unpasteurized milk commonly implicated, also thermal abuse often implicated	
Group D streptococci: *S. faecalis, S. faecium, S. durans, S. avium,* and *S. bovis*	Infection	Diarrhea, abdominal cramps, nausea, vomiting, fever, chills, dizziness	Various, but underprocessing or poor sanitation often implicated	
Vibrio cholerae non-O1	Infection	Diarrhea, fever	Shellfish from contaminated waters	
Vibrio cholerae O1	Infection	Diarrhea (severe, watery)	Shellfish from contaminated waters	
Vibrio parahaemolyticus and other vibrios	Infection	Diarrhea, vomiting, headache, fever, and chills	Raw, undercooked, or cooked but recontaminated fish, and shellfish	Halophilic
Vibrio vulnificus	Infection	Healthy: gastroenteritis; susceptible: septicemia	Shellfish, crustaceans	Halophilic
Yersinia enterocolitica	Infection	Diarrhea, vomiting, headache	Meats, oysters, fish, and raw milk	

- infection (colonization of sites within the host's body or cells),
- intoxication (production of toxins by bacteria in foods prior to them being consumed), and
- toxicoinfection (production of toxins in the consumer's body after an infection is established).

For most food-borne pathogenic bacteria, knowledge of their ecology in foods will be essential to the estimation of risk. This is because the risk to the consumer will usually be closely related to the number of cells ingested, whether through the amount of toxin formed in foods or through the probability of infection.

Dose-response models describing parenteral infections relate the probability of infection or illness to the number of microorganisms ingested. This interpretation differs somewhat from dose-response information for chemical toxins in which the severity of the disease outcome (i.e., the response) is related to the dose. In current risk assessment approaches the severity of the response is not usually considered discretely, although there is some evidence that incubation period and disease severity can be dependent on the dose ingested. Instead, the dose-response relationship has usually been used to estimate the probability that an infection will result from a particular dose given the variability in consumer response and variations in pathogen virulence according to strain, history, and food. In some risk assessments, the probability of overt disease is modeled as being conditional on the probability of infection. Even so, disease severity has rarely been explicitly modeled in microbial food safety risk assessments to date. Many dose-responses assume that even one infectious particle has the potential to cause infection. There is often an asymptotic dose, i.e., beyond which the probability of infection/illness does not increase further. The exponential model is the simplest dose-response model used in microbial risk assessment and predicts a direct proportionality between dose and risk of illness below the asymptotic ("saturation") dose. The beta-Poisson and hypergeometric models also predict a direct proportionality between dose and risk of infection in the low-dose region. Such models embody the concepts of single-hit theory for infection, and independent action of pathogens, and are consistent with the recommendations of a WHO/FAO Expert Consultation (2003), which regarded those concepts as "scientifically most plausible and defensible." Models such as the Gompertz or Weibull-Gamma predict more complex dose-response relationships in the low-dose region. They have also been reported to fit the available data equally well.

Each of the organisms in Table 2 is a distinct species and, given the diversity of the *Bacteria*, their individual characteristics and ecology vary widely. Preferred temperature, pH, and salinity ranges for growth vary, as do ranges of tolerance to other environmental conditions and the mechanisms of pathogenesis. For example,

- Organisms in the genus *Clostridium* and *Bacillus* are capable of producing endospores. Endopsores are an extremely resistant, metabolically inert, resting stage. To understand the ecology of spore-forming bacteria in foods, it is important to understand the limits of tolerance of both the spore form and the active (vegetative) form and the environmental triggers for interconversion between them.
- *Listeria monocytogenes* is adapted to growth at cooler temperatures and represents a hazard in refrigerated foods with long shelf lives (e.g., >2 weeks). Conversely, *Clostridium perfringens* is adapted to warmer termperatures and grows very rapidly at 45 to 50°C. In general, it is regarded as a problem in foods that are held hot or are not cooled quickly enough.
- *Vibrio* spp. are marine organisms and are almost exclusively associated with seafoods.
- Within each species some variation will occur in limits to growth and rates of growth and potentially virulence. *Bacillus cereus*, as a group, displays a particularly wide variation in growth limits and pathogenicity between strains of the species. Similarly, there are several distinguishable types of pathogenic *Escherichia coli* that cause distinct disease syndromes. These differences arise from genes carried on "plasmids," which are extrachromosomal genetic elements that can be exchanged between cells of a species or within closely related species. Plasmids are not required by the cell for growth and survival, but they contain genes that, when expressed, may change characteristics of the host that affect risk, e.g., virulence, resistance to stresses (including antibiotics), and often confer competitive advantages to bacteria in some situations.

These differences can be important when estimating the risk from pathogenic bacteria in food, in particular, under conditions that are marginal for growth or for survival. As food processors and manufacturers are increasingly attempting to find the mildest conditions that provide the required shelf life and integrity of foods, this variability is likely to assume increasing importance in risk assessments.

MICROBIAL ECOLOGY OF FOODS

When present in food pathogens may increase, decrease, or remain constant in number (or concentration). The conditions are described as growth (i.e., proliferation), death (or "inactivation"), and stasis (or survival), respectively.

(The term "inactivation," rather than death, is commonly used in food microbiology because, with currently available methods, it is impossible to determine whether a cell is alive, only that it cannot reproduce. Thus, the term inactivation is used to reflect this uncertainty. In many circumstances, if not most, inactivation is regarded as synonymous with death of the cells.) The extent to which growth and inactivation occur is governed by the specific characteristics of the organism itself and the environment that the pathogen experiences in the food or food-processing environment.

Guidelines for microbiological risk assessment advocated by the Codex Alimentarius Commission (Codex, 1999) state that the risk assessment should explicitly consider the dynamics of microbiological growth, survival, and death in foods. (The Codex Alimentarius Commission is an international organization created in 1963 by the United Nations Food and Agriculture Organization and World Health Organization to develop food standards, guidelines, and related texts such as codes of practice under the Joint FAO/WHO Food Standards Programme. Those standards and guidelines govern the safety standards for food in international trade.) Those dynamics are considered in brief below.

Patterns of Microbial Population Change over Time

Time is a key determinant of the fate of pathogens in foods. Microbial growth is characterized by a *rate* of population increase. Microbial death for an individual cell is a stochastic function of time or, when considered at the level of cell populations, can be characterized as a *rate* of population decrease. Thus, populations of microorganisms can both increase and decrease in foods during the life of the product as it moves along the farm-to-fork chain. Due to the partitioning of microbes among units or batches of foods, the *sequence* of environmental conditions can also be important. Lethal conditions (e.g., a heating step in food processing) followed by growth-permissive conditions can lead to a vastly different risk than the equivalent conditions applied in the reverse order. For example, if the lethal conditions completely eliminate all viable cells in a batch or unit of food, subsequent growth is not possible and, unless recontamination occurs, the risk to the consumer is zero. Alternatively, if growth is allowed prior to the lethal conditions, the number of cells then present may "overwhelm" the capacity of the treatment to eliminate them. Subsequent growth of survivors, resulting in substantial increase in risk to consumers, is possible in the latter scenario.

Differences in lethality due to the sequence of application of inimical treatments have also been reported, highlighting again the potential complexity of interactions between microbial pathogens and the environmental conditions they encounter in foods.

Patterns of microbial population decline
In general, when environmental conditions alone, or in combination, preclude the growth of microbes, those microbes will die. The rate of death is affected by the severity of the environmental conditions, with temperature apparently having a dominant effect. In general, for a given set of lethal conditions (e.g., due to inimical water activity, pH, specific growth inhibitors, etc.), warmer temperatures accelerate the inactivation rate and cooler temperatures reduce it.

Historically, the kinetics of microbial inactivation are described as a first-order reaction, that is, a constant *proportion* of the population is assumed to die per unit of time. Thus, inactivation is assumed to be essentially an exponential process and, accordingly, data are usually plotted as log(survivors) versus time with a straight line of negative slope expected to result. A wealth of evidence shows, however, that the kinetics of inactivation are often more complex than can be described by simple first-order kinetics (see, e.g., Cerf, 1977). Thus, concave upward curves (suggesting a reduction in inactivation rate over time), concave downward curves (suggesting that there is a "lag" before inactivation commences), and biphasic and even triphasic (or "sigmoid") inactivation curves are observed when inactivation data are plotted semilogarithmically. Examples of the types of inactivation kinetics observed are shown in Fig. 1.

Proponents of simple exponential death kinetics have suggested that non-first-order inactivation kinetics are artifacts due to lack of experimental control. There is much evidence, and there are theoretical reasons to believe, however, that more complex kinetics are not only possible but probable, in particular, in situations where inactivation rates are relatively slow (i.e., of the order of minutes, or longer, for a 90% reduction in cell numbers).

Strict first-order inactivation kinetics require that:

- all cells in the population have identical and constant resistance to the lethal stress applied,
- inactivation results from a "hit" to a site in the cell that results in death,
- a constant probability is associated with that site being "hit" in a unit of time, and
- the probability of a lethal hit increases with the severity of the conditions.

Cold temperature may be a special case because increasingly cold temperature reduces the rate of inactivation, even when another environmental factor is responsible for prevention of growth. Very cold temperature is routinely used for long-term storage of microbial cultures. However, microorganisms (other

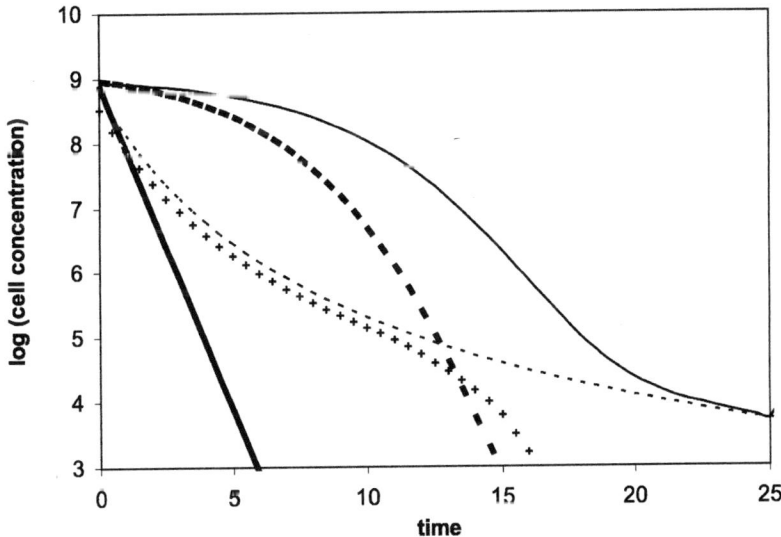

Figure 1 Patterns of microbial population inactivation kinetics. The underlying responses are believed to be log-linear (a *constant* proportion of the population is killed per unit of time) as shown by the solid heavy line, but variations in which (i) the rate of inactivation declines with time ("tailing," shown by the light dashed line) or (ii) there is an initially slow rate followed by a more rapid, log-linear inactivation "shoulders" (heavy dashed line) are also commonly reported. Yet more complex kinetics may also been seen (faint, solid line; dotted line).

than viruses) are autonomous living cells and have the ability to respond to changes in their environment. It is now known that bacteria, in particular, are able to increase their resistance to a variety of lethal stresses during prior exposure to nonlethal stress. This has been termed "habituation." It is also plausible that even cells that are in a lethal environment may be able to modify their physiology to increase their resistance if they are able, by chance, to avoid a lethal "hit" for long enough. Moreover, within any population of cells it is reasonable to expect that there will be a distribution of physiological states, a function of the conditions that individual cells have experienced both prior to contaminating the food and while part of the food (note: at the scale of a microbe foods are not necessarily homogeneous and may present a variety of microhabitats both in time and position).

A possible reason for reluctance to accept non-first-order inactivation kinetics is that the assumption of simple exponential decline greatly simplifies the calculation of the extent of inactivation, e.g., for the purposes of designing thermal food processes to achieve a required level of microbial inactivation or to assess the relative efficacy of different thermal processes. Specifically, under the assumption of first-order inactivation kinetics, the rate of inactivation under defined constant conditions is characterized by the D-value, which is the time required for 90% of the population to be inactivated. In thermal processing, the z-value is used to characterize the rate of change of D-value as a function of temperature, though the concept could, in principle, be applied to other methods of microbial inactivation, such as high pressure or acidity. The z-value is the increase in temperature required to reduce the D-value by a factor of 10. Typical z-values for bacteria are in the range of 5 to 10°C. If inactivation were not strictly exponential, however, the D-value would also become a function of treatment time (or, perhaps, population density). This would considerably increase the complexity of predicting the antimicrobial efficacy of thermal processes, e.g., for the design, optimization, or evaluation of such processes.

The notion that microbial inactivation is a stochastic process is an important consideration in microbial food safety risk modeling because, when modeling inactivation, survivor levels of less than one cell per unit of food are frequently predicted. Clearly, one cannot have a fraction of a cell, and it is tempting to regard these values as equivalent to no viable cells per unit (e.g., can, bottle, pouch) of food. The correct interpretation of the "fractional cell," however, relies on recognizing that death/survival models are based on the *probability* of a cell surviving the process. Thus, if the modeling predicts that a population of 10^3 cells in a unit of food is reduced a hundredfold (i.e., a "2D" inactivation process), each cell has a 1% chance of surviving the process. Accordingly, approximately 10 cells from the original 1,000 are expected to survive the process and remain viable in each unit of the food. If, however, the initial population was one cell per unit of food and the same treatment was applied, we would still expect that each cell had only a 1% chance of survival and, prima facie, that no unit was likely to contain a viable pathogen. Considering this in another way, however, if the contents of 1,000 units were combined, 10 viable cells would be expected to remain among the 1,000 units after the treatment. In other words, among 100 units initially contaminated at a level of one cell per unit, one unit (on average) would be expected to contain a viable cell after the treatment. Thus, at increasingly low contamination levels, the effect of an inactivation process changes from consideration of *concentration* of pathogens per unit of food to consideration of *frequency* (or probability) of contamination of units of food. This issue can be regarded as an example of "partitioning," considered as one of the basic processes affecting pathogen contamination level and frequency in foods (see elsewhere in this chapter and the chapter by Nauta [chapter 4], in which methods for modeling partitioning are described).

Patterns of microbial population growth

Microbial growth in individual units of foods is effectively a batch culture and often can be described by the classical population growth curve (Fig. 2), in which the logarithm of the number of cells is plotted against time because microbial populations increase exponentially with time.

The curve consists of four phases termed lag, exponential, stationary, and decline, of which the lag phase and growth rate during exponential phase have been most studied. The curve is often considered to be sigmoid because the decline phase is of little relevance in food microbiology and often ignored.

Lag times, during which no growth occurs, can be observed when bacteria are transferred to a new environment, such as from processing equipment, or water, or human skin, onto food. The lag time is usually interpreted as a period of adjustment to the new environment, during which biochemical and physiological changes are undertaken by the cell to enable optimal exploitation of the new environment. In some situations, e.g., if cells are transferred from an environment in which they are actively growing to another similar environment, no lag occurs.

During the exponential phase, the cells in the population are operating at near-optimum metabolic efficiency, with the sole aim of producing more

Figure 2 Microbial population growth curve in a closed system, e.g., a unit of food, showing the generally recognized stages of "batch" population growth.

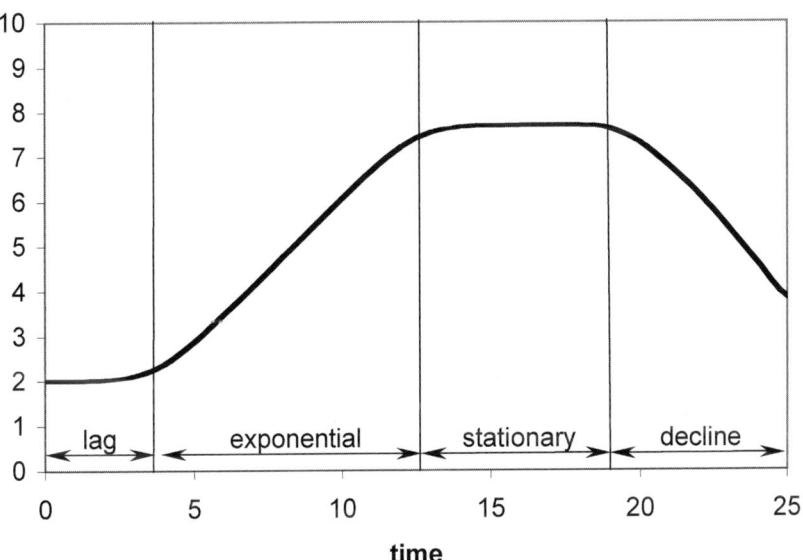

copies of themselves as rapidly as possible. The exponential or "specific" growth rate (i.e., the number of *multiplications* per unit time), μ, is given by:

$$\mu = \left(\frac{\partial N}{N \partial t}\right)$$

where N = number of cells and t = time, and can be determined from the slope of the relationship between time and ln(N). The generation time is the time taken for the number of cells to double, i.e., for $\partial N = N$, ∂t corresponds to the generation time. While strongly affected by environmental conditions, the exponential growth rate of a specific microorganism in a defined environment is highly reproducible.

Unless affected by the presence of high numbers of other microorganisms, exponential growth continues until some maximum population density (MPD) is neared. At that point population growth slows and eventually ceases. This is called the stationary phase. If the population remains in the same, closed environment, the population density will begin to decline again, described as the "death" or "decline" phase. The number of cells at stationary phase (i.e., MPD) for a specific bacterium is largely independent of the abiotic environment except under combinations of environmental conditions that nearly prevent growth. For many organisms and many foods the MPD is in the range 10^9 to 10^{10} cells per gram or ml, or cm^2, of food, although this level can decrease when the organism is in a food that strongly reduces growth rate, i.e., product formulations that are near the limits of the microbe's growth range.

Several models to describe the sigmoid population growth curve have been proposed (for reviews see McMeekin et al., 1993; Ross and Dalgaard, 2003). Two of those models have gained prominence in food microbiology, namely an empirical modification of the Gompertz function and reparameterizations of it, and a development of the "logistic," or "autocatalytic," function that has become known as the D-model, or the "Baranyi model."

The boundary between growth and death

An important principle in the modern food industry is that when several factors inhibitory to microbial growth are applied in combination they can often prevent microbial growth at levels lower than any one of the factors acting alone. This has been termed the "hurdle" concept, and its practical application named "hurdle technology" or "multiple barrier" technology. In short, when one factor in a microbe's environment is suboptimal for growth, it reduces the growth range of the organism for other environmental factors. The mechanisms of that interaction are not known in detail.

The combination of levels of factors that either individually or collectively *just* prevent the growth of a microorganism is imagined to form a smooth boundary in multidimensional space. An example, for two variables, is shown diagrammatically in Fig. 3a and an example based on actual data is in Fig. 3b. As each environmental factor becomes less optimal for growth, the growth rate is reduced until, ultimately, growth is prevented. When the value of one environmental factor is extreme, but still permits growth, other factors must be at, or near to, their optimal levels for growth to occur. For example, McMeekin et al. (1987) demonstrated that at very high salt concentrations (a_w = 0.826) the halophilic bacterium *Staphylococcus xylosus* could only grow at temperatures above 28°C, while at optimal salt levels (a_w = 0.889) growth was possible at temperatures as low as 8°C.

The transition between growth and no growth has been observed to be abrupt in many cases, and the relationship between environmental conditions and growth rate often becomes discontinuous when the growth limit is reached, i.e., at these combinations of conditions growth rate declines abruptly to zero for a very small change in the environmental conditions. This phenomenon has significance for risk assessment, in particular, of foods with a long shelf life, because while growth is slow at conditions that just allow growth, significant increases in cell numbers can occur over time. If conditions (i.e., preservative levels in product, storage temperature, etc.) are just slightly more extreme, however, no growth will occur no matter how much time is allowed. Moreover, microbial inactivation and decreased risk might ensue. Thus, the risk estimate can change by a large amount for a small change in predictor variables, in particular, when amplified by the effects of time. Such behavior is one of the characteristics of a chaotic system. When predicting growth of microorganisms using predictive microbiology models, risk assessors must also consider whether growth would *ever* be possible given the composition of the product and the conditions of its storage. If growth is possible, a growth rate model is required to estimate the extent of growth. Equally, if growth is not possible, the inactivation over time may be able to be estimated by recourse to a mathematical model describing microbial inactivation rate.

Microbial Ecology of Food Made Easy

Most microorganisms are living and capable of independent existence. As such, they can recognize and respond to changes in their environment, including increasing their resistance to the effects of a hostile environment, and can also modify their environment in some circumstances. Thus, there is a range of consequences to the cell upon exposure to a specific environment, adding dimensions of complexity, uncertainty, and variability to modeling

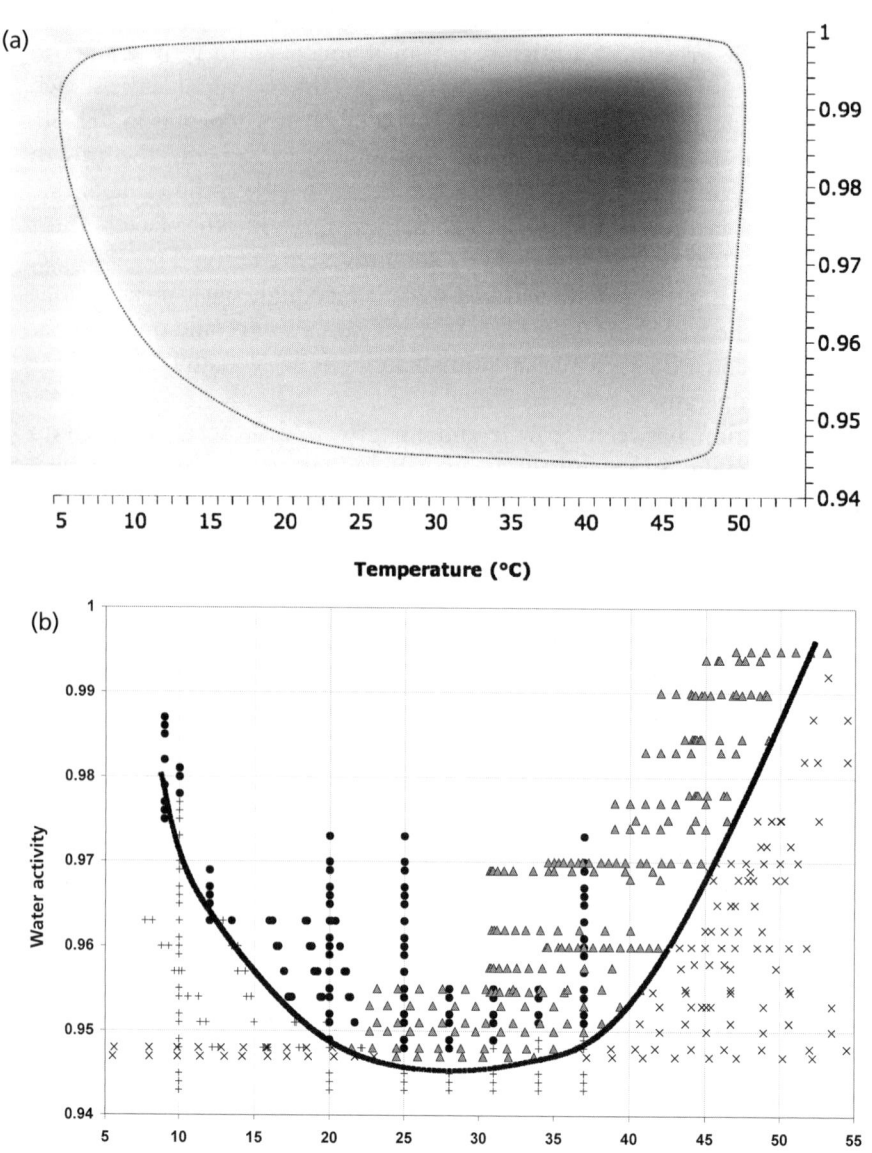

Figure 3 (a) Diagram illustrating the effects on microbial growth rate of two factors (temperature and water activity). The figure also illustrates the concept of the growth/no-growth boundary (dotted line) where the combination of factors at suboptimal levels eventually prevents growth. In the figure, the darker shaded areas represent regions of faster growth, while lighter shaded regions represent regions of slower growth. The rate of change of growth rate due to environmental conditions is not a constant, e.g., growth rate declines more quickly at temperatures above the optimum for growth than below. (b) An example of experimental data describing the water activity and temperature growth/no-growth interface for *E. coli*.

the fate of pathogens and consequent risk compared with other food-borne hazards.

Environmental factors that affect the fate of pathogens in foods include both abiotic (nonliving) and biotic components. Among the abiotic factors are temperature, levels of oxygen and other gases, osmotic pressure (due to addition of solutes such as sugars or salt, or removal of water), acidity, nutrient levels, and the presence of specific preservatives such as nitrite, organic acids, smoke components (phenols), and so on. Each of these factors can inhibit or prevent the growth of pathogens, but the levels required to prevent growth are not always acceptable in food for aesthetic or safety reasons, e.g., at high levels some additives may be toxic to consumers. As noted above, those factors can act additively, and sometimes synergistically, to retard growth. If growth is possible, then the population will increase, as shown in Fig. 2, at a rate governed by the combined inhibitory effects of the environmental conditions and the characteristics of the organism itself.

The presence of other microorganisms in foods (the "biotic" environment) can also affect the fate of pathogens in foods through competition for resources or production of inhibitory compounds. Food-processing operations, too, can affect the fate of pathogens as can packaging systems. It is important that risk assessors recognize the influence of these factors and how they can change in the product as it moves along the farm-to-fork chain. Many schemes to assist in understanding of the microbial ecology of foods have been proposed and can help to reduce, or at least organize, the apparent complexity.

For the purposes of microbial food safety risk assessment, Nauta (2002) proposed that six basic processes can affect the number of pathogens in foods as they move along the farm-to-fork chain. These processes are:

- growth (as discussed above),
- inactivation (as discussed above),
- mixing (combining food ingredients, with the possible result that larger amounts of food may be contaminated from their individual components but with no net change in the total number of microbes in the batch),
- partitioning (subdivision of batches into smaller units, with the result that frequency of contamination may be reduced but with no net change in the total number of microbes in the batch),

---

Filled symbols denote conditions under which growth was observed, while crosses and plus symbols indicate conditions under which growth was never observed. Different symbols indicate data derived from two separate modeling studies in this laboratory but show a high level of consistency, although some inconsistent data are apparent. The dotted line is empirically derived and included simply to highlight the interface.

- removal of microbial contaminants (by various physical processes, e.g., separation of components or transfer to other materials, so that the level and/or frequency of contamination in the food decreases overall), and
- cross-contamination (with the result that the frequency and/or level of contamination in the food increases overall).

Nauta reasoned that development of mathematical descriptions of each of these fundamental processes would greatly simplify the conduct of microbiological risk assessments by enabling risk assessors to incorporate data specific to a product/process risk to existing model structures that had attained general endorsement by the scientific community. The approach, termed Modular Process Risk Modeling, is considered in detail by Nauta in chapter 4.

Prior to the application of mathematical modeling techniques to food microbiology Mossel and Ingram (1955) proposed that factors affecting the microbial ecology of foods could conveniently be considered under four headings, a classification that remains in use. Those categories are:

- intrinsic factors
- extrinsic factors
- implicit factors
- processing factors

which are explained in greater detail below, and from which elements analogous to Nauta's categorization are apparent.

While these categorizations per se do not provide information about the microbial ecology of foods, they offer systematic frameworks for identification of factors that could affect the frequency and level of contamination of pathogens in foods. As such, they can be useful aids for developing conceptual models for microbial risk assessment of foods that encompass the influence of microbial ecology and physiology.

Intrinsic factors

In Mossel and Ingram's scheme, intrinsic factors are the physicochemical properties of the food itself such as structure, water activity, acidity, or the presence of specific organic acids, specific nutrients, etc. For example, lactic acid levels can increase in animals due to stress prior to slaughter (or catch, in the case of fish) and can lead to significant reduction in the pH of the flesh. The increased lactic acid and pH act synergistically to inhibit microbial growth.

Some organisms have specific nutritional requirements. For example, at chill temperatures *L. monocytogenes* requires certain amino acids to be available preformed and free in its environment. Milk does not contain high levels of free amino acids because most are present in the form of proteins. (Note: this has been suggested as a reason why *L. monocytogenes*, which requires certain amino acids to be available in its environment to enable its growth, does not grow as well in milk as in some other foods; conversely, it has been shown that other organisms often present in milk can break down the available proteins, releasing free amino acids and promoting the growth of *L. monocytogenes*.) Other natural antimicrobial compounds or systems exist in some foods, e.g., the lactoperoxidase system in milk (the effect of which is diminished, however, through pasteurization, which is a "processing" factor). The physical structure of some foods can also affect their microbial ecology. For example, cuts of meat can provide two quite different environments to contaminating microbes. The first is on the outside of the intact muscle bundle contained within the fascia. This region can become partly dry during storage, thereby reducing the water activity and inhibiting or preventing growth of many types of gram-negative bacteria. If the muscle bundle is cut, however, individual muscle cells are ruptured, releasing water and nutrients and increasing the potential for microbial proliferation.

The skins/peels of fruits and vegetables are an example of a simple physical barrier that protects the more nutritious inner parts of the fruit from microbial attack and degradation. Processing may alter the intrinsic properties of the food, in particular, from the perspective of a microorganism, if it breaks down physical barriers to microbial invasion, causes mixing of components of the foods, adds or removes water, and so on.

Extrinsic factors

The most obvious example of an extrinsic factor is temperature. Both heating ("hot holding") and refrigeration are widely used to prolong the storage life of food and to minimize the risk of microbial food-borne disease. Food processors can also manipulate the gas mixture surrounding a food, as is done in modified atmosphere packaging (using mixtures of CO_2, nitrogen, and oxygen, or perhaps carbon monoxide). High concentrations of carbon dioxide inhibit microbial growth, and reduced oxygen availability greatly reduces the growth of obligate aerobic microorganisms, e.g., psychrotrophic pseudomonads that are characteristically associated with the spoilage of moist, refrigerated, proteinaceous foods, or molds that are common airborne contaminants and that spoil drier products (e.g., cheese) unless oxygen is excluded from their packaging.

"Biotic" factors refer to the effects of other microorganisms in foods that can interact with, and affect, the ecology of the organisms of primary concern. Potential effects of spoilage and other microbes on food-borne pathogens are considered in greater detail elsewhere in this chapter.

For some foods (e.g., fruits), the relative humidity of storage areas is exploited to preserve the integrity of the product and so extend shelf life. In doing so, growth of toxigenic molds can also be inhibited. Relative humidity control is also an important part of production of some foods (e.g., fermented meats) and is required to achieve a desired final water activity that, in conjunction with reduced pH, inhibits microbial growth.

Implicit factors

Implicit factors are the characteristics of the microbes themselves, such as the ranges of pH, water activity, oxygen, and temperature over which they can grow, the maximum rates at which they grow under various conditions, and their resistance to other food preservatives. A wide diversity of food-borne microbial pathogens exists and the responses of those pathogens to environmental factors can differ widely between species. For example, some microbes are specifically adapted to high temperature, others are adapted to low temperature, some can tolerate highly acidic or hyperosmotic conditions, while others are rapidly inactivated in such environments. Even within a single species, there can be important variation in the effects of environmental conditions on microbial growth or inactivation. The virulence of microbial pathogens also varies between "strains," i.e., cells that are members of a single species, and in some cases as a function of environmental conditions. Implicit characteristics of the specific pathogen of concern, and their variability between strains, should be considered as part of risk assessments.

Processing factors

Processing factors include actions performed during processing and distribution that change the food and thus change the ecology of the product as well. For example, slicing breaks down the structure of some foods, releasing nutrients and moisture and increasing the potential for microbial growth. Additionally, slicing involves direct contact of the cutting blade surface with the food. Blades become covered in food substrates and nutrients and contaminating bacteria can grow in such niches if they are not properly and frequently cleaned. Thus, slicing can introduce microbial contamination to foods, or to parts of foods, that were previously sterile. Brine injectors used for processing of some types of meat and fish can also introduce bacteria to otherwise sterile sites in foods and from which they may be more difficult to eliminate by normal food-processing techniques.

Washing is intended to remove microbial contaminants but may introduce other contaminants if the wash water itself is not kept free of microbial contamination, in particular, if the water is recycled or used for many batches of product. Heating, irradiation, and other physical treatments that inactivate microbes alter the composition of the "microbiota," i.e., the totality of microbes present on foods, usually greatly reducing the population of non-spore-forming bacteria. In doing so, they can increase the chance that spore-forming organisms, potentially including food-borne pathogens (see Table 2), will subsequently germinate and become dominant on those foods. As noted, the "natural microbiota" of a food can affect the proliferation of food-borne pathogens.

DATA NEEDS AND SOURCES

From the preceding discussion it is apparent that many types of data might be required to predict the fate of pathogens in foods from "farm to fork." While this chapter is focused on microbial ecology of foods, many other types of data are also needed to fully characterize consumer exposure to a specific microbial food-borne hazard. Those data, their sources, and their appropriate use in "exposure assessment" were the subject of an FAO/WHO expert consultation that prepared a set of guidelines for the conduct of exposure assessment in the context of microbial food safety risk assessment (FAO/WHO, in preparation). This section considers sources of data that facilitate description and prediction of the microbial ecology of pathogens in foods and food ingredients from "paddock to plate."

As noted earlier, the risk to consumers ultimately depends on the number of pathogens consumed and the frequency of exposure. To estimate the dose ingested it is necessary to have knowledge of:

- the concentration of the pathogen at some earlier point in the product life, e.g., numbers of the pathogen initially present on the food, or ingredients used to make the food, and
- the change(s) in the number or concentration between the point at which the concentration is known and the time of consumption.

The latter is a function of the implicit characteristics of the pathogen and:

- its lag time in the food (or in the case of spores, times for germination and outgrowth),
- its potential for growth,
- whether conditions or processes cause inactivation, and
- whether further contamination of the product occurs prior to consumption (i.e., cross-contamination).

The potential for growth and inactivation, in turn, depends on intrinsic, extrinsic, and processing factors including:

- the chemical composition and physical structure of the food,

- the time(s) and condition(s) of storage or treatment and, in some cases,
- the presence and types of bacteria or other microbes that grow in the food.

Data on Explicit Factors

Information on processing and intrinsic and extrinsic factors can often be obtained by expert solicitation and from company records such as Hazard Analysis, Critical Control Points (HACCP) plans and records, product recipes, or finished product parameters. Data relating to temperatures during distribution, storage and retail display, and storage in domestic refrigerators are more difficult to obtain although several studies have been undertaken and published. Data describing the time spent in each stage of the farm-to-fork chain are not usually readily documented but can be deduced from expert solicitation and knowledge of product shelf lives, together with rates of production, retail stock records, and so on. Information on distributions of times to consumption can also be gleaned by surveys of product at retail sale that assess the remaining shelf life of product still available for sale by comparing the date of production and/or "use by" or "best before" dates with the survey date.

Estimating Initial Microbial Loads

Data concerning initial levels of pathogens in foods are also hard to obtain because they are not routinely assessed except by the largest food businesses. To date, such data have usually been derived from collations of published studies. These include large, coordinated "snapshot" or "baseline" surveys conducted by national governments or industry organizations and small, ad hoc surveys by independent researchers. Some government organizations, however, may also have useful information arising from monitoring programs. Knowledge of levels of *spoilage* bacteria in foods can also be important in predicting the potential for growth of pathogens and may be useful to generate better estimates of risk. While food processors often collect data on the total microbial load (i.e., "total viable count" or "standard plate count") on foods at the time of manufacture, those methods do not differentiate and enumerate the organisms that will eventually cause spoilage of the product. Instead, those organisms (sometimes called specific spoilage organisms or SSOs) are usually a subpopulation of the initial microbial load. Nonetheless, variations in total bacterial load may be indirect indicators of plant/process hygiene, and increased counts are likely to be correlated with increased levels of SSOs and reduced shelf life.

Quantification of the lag time of pathogens in foods is less certain again. While many published studies do report microbial *population* lag times, lag times are widely reported to be more variable than growth rate under the same conditions. This is probably due to confounding effects of physiological history of the cell (which will dictate the amount of adjustment the cell has to perform to optimize its growth in the food) and environment, which dictates the rate at which those adjustments can be made. Few studies that report lag times include sufficient data to enable discrimination of the relative contributions of these effects.

Lag Time Information

The question of the why lag times are more variable than generation times led to the suggestion that in any population of cells there is a distribution of states of physiological "readiness to commence growth" among the cells comprising that population. Some cells were envisioned to be more ready to commence replication in the new environment and would be expected to have shorter lag times while other, less ready cells would be expected to have longer lag times. It was further envisaged that the lag time of a population of cells (the property that is traditionally measured) would be related to the average physiological state of all the cells present but influenced more heavily by the cells with the shortest lag times. Those cells would be expected to initiate growth first, with their progeny comprising a larger proportion of the population. If many cells were present in the population, many states of physiological readiness would be represented among the cells and the lag time expected to be more reproducible, but if the population comprises only a few cells, the lag time of the population would be much more dependent on the lag times of individual cells, namely, the chance of cells with very short lag times being present, or not, in the population. These predictions have been supported by several recent experimental studies on lag time variability as a function of inoculum density.

Subsequent studies developed methods to measure the lag times of individual cells, using image analysis systems to detect the time to first division, and subsequent divisions of cells, or based on the time to detection of growth of broth cultures inoculated with one or a few cells only. In the latter case variability in the time to detection was shown to be a function of the variability in the lag time of those cells, rather than due to variability in growth rate of those cells once their lag phases were resolved. Similar studies confirmed the much lower variance in generation time among growing cells.

Using those experimental approaches it has been shown that, as the stressfulness of environments increases, or as the degree of damage to cells (e.g., by heating) is increased, lag times increase but also become more variable. In addition, the distribution tends to become more asymmetrical as the lag times become relatively longer (e.g., due to more stressful conditions or cell injury). Thus, (favorable) conditions leading to short lag times tend to have skewed lag time distributions with peaks near zero but with long "right-hand tails" because lag times cannot be less than zero. Researchers have sought to find the best mathematical description of lag time distributions but with little consistency between the findings of different groups.

While for small cell populations, the number of cells at some later time significantly depends on the lag time of the individual cells originally present, the effect is much less pronounced as the number of cells initially present increases because there is a greater probability of cells with short lag times. This conclusion, however, depends on the mean lag time. For severely injured or stressed cells, in which a wider distribution of lag times is expected, variability in lag times is observed even in larger populations. Reports suggest that lag time variability, and mean population lag times, increased when initial populations are smaller than 40 to 50 cells, but that the variability in mean population lag time is greatly reduced when initial cell (or spore) populations exceed 500 to 1,000 cells.

In the context of microbial risk assessment, the significance of lag time variability also depends on the magnitude of the lag time compared with the time required for the cells initially present to reach a dose that is highly likely to cause illness. When the time required to reach levels highly likely to cause illness is large compared with the lag time, variability in lag time will be less important as a source of variability in risk to consumers. Several risk assessments have demonstrated this effect for *L. monocytogenes*, for which the ID_{50} is estimated to be very high compared with typical initial contamination levels, requiring many generations of growth before a significant probability of illness results. (Recognizing that individual susceptibility to pathogens varies and that virulence varies among pathogens, the ID_{50} is a concept that expresses the dose of pathogens that is required to establish infection and cause illness in 50% of a test population.)

While increasing variability of lag times with increasingly stressful growth environments has been widely reported, other reports have indicated that the coefficient of variation of the lag time is relatively constant, i.e., while variability in lag time increased in absolute terms, the increased variation was proportional to the increased mean lag time.

In general, the lag time is affected by environmental conditions in the same manner as generation time, i.e., less optimal conditions decrease the rate at which the lag phase is resolved by the same relative amount as the growth rate is reduced by those conditions. Thus, the ratio of lag time to generation time (relative lag time, or RLT) effectively normalizes the effect of growth environment on lag time and RLT expresses the lag time as an equivalent number of generation times in the same environment. If RLT increases, it infers that more "work" has to be done by the cell to adjust to the new environment. Analyses of frequency distributions of RLT for many bacterial pathogens derived from many independent studies suggest a pronounced peak in the range 3 to 6 under a wide range of experimental conditions. Thus, while lag time is apparently highly variable, RLT is more uniform and reproducible. RLT can also simplify the growth-modeling process because use of the RLT as a variable enables the growth rate and lag time resolution to be predicted by a single growth rate model.

Implicit Factors
Unlike the data constraints referred to above, implicit factors are reasonably well documented for many pathogens for the major factors determining growth rates. Limits to growth are also reasonably well characterized for individual factors, but less so for combined factors. There is also growing evidence, and documentation, of between-strain variability in tolerance to environmental factors.

The International Commission on Microbiological Specifications for Foods (ICMSF, 1996) provides a useful collation of relevant growth rate and inactivation rate data for a variety of food-borne pathogens under various conditions. The ComBase (www.combase.cc) database is another powerful resource for microbial risk assessment. It is a publicly accessible searchable database of >35,000 records of microbial responses over time to environmental conditions. The database can be searched by type or species of organism, type or class of food, pH, temperature, water activity (or NaCl concentration), and specific food conditions.

Another important resource in exposure assessment of microbial pathogens in foods is the predictive microbiology literature and related data and models discussed below.

Rates of Population Change: Predictive Microbiology
The "rules" governing the microbial ecology of foods are increasingly well understood and, in many cases, the fate of pathogens in foods *can* be predicted from quantitative knowledge of the biology of the pathogens, and the conditions to which they are exposed along the farm-to-fork chain, using the approach and resources of "predictive microbiology."

Predictive microbiology seeks to summarize knowledge of the ecology of food-borne microorganisms and to express that data and the general patterns of microbial responses as mathematical models. The models enable predictions to be made, by interpolation, about responses to new situations and scenarios not specifically tested. The approach is based on the premise that responses of microbial populations to environmental conditions are reproducible. By characterizing environments in terms of those factors affecting growth, survival, and inactivation, it is possible, from past observations, to predict responses under similar conditions by interpolation. Thus, in some situations, it is possible to predict microbial changes in a food by monitoring the environment of the food over time rather than relying on microbiological analysis.

The general patterns of microbial response to environments in foods are described by mathematical models of a form applicable to many types of microbe. Models specific to organisms of interest are derived by choosing optimal values for the model's parameters. This is done using regression analysis so that the model best describes the experimental data for responses of the food-borne organism or group of organisms.

Models, including those that describe growth rates, inactivation rates, and limits to growth under combined stresses ("growth/no growth interface" or "growth boundary" models), are available for a wide range of food-borne bacterial pathogens and specific spoilage organisms and a smaller number of yeasts and molds.

Models that integrate growth and death models into a single model form, predicting growth for a particular range of predictor variables and population decrease at values beyond those ranges, have also been proposed. Technical problems associated with their development have been identified, including mathematical transformations (usually required for regression analysis) when rates of change of populations are negative. A review of primary models (McKellar and Lu, 2004) suggested that the bases of such models proposed at that time were questionable (e.g., that microbial responses to environmental changes are instantaneous) and development of such models is not currently an area of strong research activity.

Model nomenclature
In the predictive microbiology literature "primary" models are those that describe the change in microbial populations over time under constant environmental conditions, e.g., population growth curves or death curves, or the probability of growth. "Secondary" models describe the influence of environmental conditions on the parameters of the primary model, such as lag time, exponential growth rate, or D-value. "Tertiary" models are computer software systems that make the predictions of models readily accessible to users. From the perspective of risk assessors, most interest will center on secondary models in combination with primary models. Secondary models can be embedded within primary models to describe the effect of environmental conditions on the values of the parameters of the primary model (e.g., growth rate, D-value, lag time, etc.) and, thus, to translate information about environmental

conditions into predictions of the size of microbial populations over time, whether for growth or inactivation processes. Growth/no-growth models might also be used to assess whether the various environmental conditions along the farm-to-fork chain are expected to lead to proliferation, or inactivation, or stasis of the microbial population of concern.

Sources of secondary models and data
Models to predict growth rate, death rate, and the probability of growth under defined environmental conditions are available in the published literature. Existing models were collated and summarized by Ross and Dalgaard (2004). Most of the existing models describe the behavior of pathogenic bacteria or spoilage bacteria, but models for growth of *Aspergillus* spp. and various spoilage yeasts and molds are also available. Most data for inactivation of pathogens and spoilage organisms relate to thermal inactivation. Relatively few models describing the effects of multiple factors on the rate of inactivation are available.

The suitability of existing predictive microbiology models for risk assessment has been questioned, in part, because they are usually presented as deterministic functions rather than explicitly considering the variability in predictions or because some earlier models were developed to predict growth under static conditions. Many models do include estimates of the error around the prediction: most regression modeling approaches require that the fitting process is done by using a response variable for which the error is normally distributed. Thus, for example, predictive microbiology models may predict the logarithm of growth rate, or square root of growth rate, because the error (including both biological and methodological sources of variability) is homogeneous in these transformations. When using such models to predict growth rates, or generation times, and so on, the effect of the backtransformation of the response variable on the error distribution must also be considered. The error estimates of models can contribute to estimation of variability in the predicted changes in microbial populations, and risk of illness, at the point of consumption.

A desideratum in risk assessment is to "let the data speak." The ComBase (www.combase.cc) database, referred to earlier, provides another means of characterizing variability in microbial responses for stochastic risk assessment models because it presents data rather than models derived from them.

Application of Predictive Microbiology Models to Risk Assessment
As noted, various authors have commented that predictive microbiology models, as traditionally developed, are not ideal for integration into risk assessment models. Most models have been developed using only one, or a

limited number of strains, of the pathogen of concern. When multiple strains have been used, they have been studied as a cocktail, with no differentiation of the behavior of individual strains, i.e., information on strain variability is not explicitly gathered. Others have noted that models are usually developed from data gathered from simplified model systems (e.g., usually broths and in the absence of competing organisms) and have questioned the relevance of such models to foods. Most models are developed from studies under constant environmental conditions, while microbial hazards in foods clearly experience fluctuating conditions. Currently, virtually all predictive microbiology models are empirical but are used within mechanistic frameworks describing microbial ecology. While more mechanistic predictive microbiology models are preferable, these are often more complex, requiring measurement or estimation of more parameters. However, greater model complexity is a disadvantage in stochastic risk assessments.

Strain variability

As noted, most predictive microbiology models currently available provide only a limited consideration of variability in responses between strains of pathogenic species of interest, with many models having been developed using a single strain. Ideally, the behavior of this strain has been compared with other strains to assess that the strain is "representative" or represents a "worst case." In other cases, models are developed using "cocktails" of strains, with the idea that under any conditions the growth or survival of the most resistant strain will dominate the observed responses. In short, most predictive models have been developed to be deliberately conservative, or "fail-safe," rather than to represent the distribution of responses among all strains of the microbial hazard of interest.

Several stochastic modeling studies have reported that variability between strains of the same species is larger than the variability due to experimental error observed in studies with single strains or cocktails of strains and that this variability can have profound effects on the outcomes of risk assessments. Variation in growth rates, inactivation rates, growth limits, and virulence between different strains of the same species has been documented, including systematic comparisons of large numbers of strains. Differences in responses are often more pronounced under conditions that are marginal for microbial growth, but these, of course, are usually the conditions of most relevance and interest to microbial food safety risk managers. From studies that have directly compared responses of different strains under identical, controlled conditions, differences in lag times between strains can range from approximately 5- to 30-fold, while differences in growth rates between strains range from 1.5- to 3-fold. Differences in D-values of *B. cereus* have been reported to vary by 5- to 30-fold for either mesophilic strains or psychrotrophic strains, with psychrotrophic strains having a lower mean D-value.

Food structure
As noted, at the scale of a microorganism, foods may contain many distinct microenvironments that can induce differences in microbial growth rate and potential. In addition, while liquid media will disperse metabolites produced by the bacteria and fungi, in solid or semisolid foods, diffusion of chemicals such as metabolites or nutrients may be constrained. Toxic metabolites may accumulate to high concentrations in the vicinity of the cell and cause growth inhibition despite that the overall density of cells in the food is low. Similarly, in a more structured food matrix, a small number of microbes may exhaust the nutrient supply in their immediate vicinity, despite that the overall concentration in the food remains high. Published studies indicate that this can occur in practice.

The most obvious example of this effect is in water-in-oil emulsions, such as mayonnaise or margarine. Emulsions are composed of droplets of water suspended in oil (or fat) or globules of oil suspended in water. In either type, microbes can grow in the aqueous phase, but not in the oil phase. If a low level of contaminants is present in a water-in-oil emulsion, only some of the water droplets will contain pathogens. The pathogens cannot migrate between droplets, and once they have fully exploited the resources in the water droplet, they cannot grow further. Thus, the total growth possible in the product is limited to the amount of growth possible in the contaminated droplets. Conversely, in the oil-in-water emulsion, a single cell initially present anywhere in the water phase has the potential to utilize all the space and nutrients in the aqueous phase because the aqueous phase is continuous throughout the product. Similarly, in many foods that permit microbial growth, a continuous aqueous phase disperses both nutrients and microbial metabolic wastes and, in such products, predictive models can provide realistic predictions of growth. Other foods, such as cheese, may present a combination of environments such that microbes may be present and grow planktonically in pockets of water and solutes or be immobilized in gelled regions. This creates difficulties in characterizing the potential for growth due to uncertainty about the actual environment that the microbes experience. To continue the example, if a contaminating cell is present in the oil, growth is unlikely to occur because the cell needs water for metabolism. Thus, at low levels of contamination there is also a stochastic element that influences the microbiological fate of the product—whether the few contaminants present are contained in fat/oil and are therefore unable to proliferate, or whether they are in the aqueous phase and able to exploit that environment.

The effects of food structure can be compounded by interactions of different microbes in proximity in a food system that immobilizes them or that inhibits diffusion of nutrients and metabolites. An additional aspect to the influence of food structure is that the microbes can be induced to enter stationary phase at much lower cell densities than would occur in liquid culture (typically ~10^9 to 10^{10} cells ml^{-1}). A range of physiological changes in cells occurs when they are stressed or enter stationary phase, including increases in virulence. This aspect is considered further elsewhere in this chapter.

A succinct review and analysis of the effects of food structure on microbial responses are provided by Wilson et al. (2002).

Microbial interactions
Numerous studies report the interactions between organisms cocultured in aqueous systems and demonstrate that effects, and inhibition, do occur at high concentrations. Where the structure of the food immobilizes cells and inhibits nutrients, these interactions can occur at much lower cell densities, provided that the cells are in relatively close proximity (described as "propinquity"). Although lower cell densities suggest that the probability of propinquity would be low, microbial cells may preferentially partition into aqueous environments, increasing their local concentration.

Environmental fluctuations
Studies in the 1960s indicated that abrupt changes in temperature, within the range that permits growth, could induce growth lags in exponentially growing cultures before growth recommenced at the rate characteristic of the new temperature. The effect was found to be dependent on the magnitude of the temperature shift and also whether the shift in temperature extended beyond a mid-range of temperature for growth, described as the "normal physiological range." It was proposed that some temperature shifts, in particular, from within the normal physiological range to temperatures beyond it, required the cells to alter the chemical composition of its membranes, requiring synthesis of different fatty acids, and to synthesize other enzymes better adapted to function at the new temperature. The induced lag time represented the time taken to complete the necessary physicochemical changes to the cells' composition. This interpretation is supported by experimental observations. More recent studies have shown that lags in growing populations can also be induced by environmental changes other than temperature.

Mild environmental stress can also induce many organisms to increase their resistance to subsequent lethal treatments, enabling cell populations to survive for longer. Increased resistance to lethal acidity subsequent to exposure to a milder, nonlethal level has been a major area of research. The induction of increased heat resistance among bacteria that are slowly heated has also been reported, and cross-protection by one mild stress to subsequent lethal levels of another stress is also widely reported. As discussed elsewhere, concave upward inactivation curves have been interpreted as indicating a distribution of innate resistance to the treatment, or the induction of greater resistance to the treatment in survivors over time. The physiological basis of these changes is considered in greater detail elsewhere. In short, however, the environmental history of an organism can have a large effect on D-values, with exponentially growing cells typically having much lower resistance, but with periods of mild stress increasing subsequent resistance. As a rule of thumb, the effect of stress- or stationary-phase-induced responses results in D-values becoming 2 to 3 times longer than in nonhabituated, exponential phase cells, though smaller and larger differences have been reported. Given that the D-value is the time taken for a 90% decrease in the population of viable cells, and that risk is proportional to absolute cell numbers, not their logarithm, relatively small changes in D-value can have a profound effect on the estimated risk. As such, caution should be exercised that D-values used in risk assessment models be appropriate to the physiological state of the cells whose inactivation is being modeled.

Model complexity
van Gerwen and Zwietering (1998) offer good discussion and practical advice on the selection of microbial growth and inactivation models for use within quantitative risk assessments. This advice formed part of an overall strategy for a stepwise development of the risk model that first focuses attention on the main determinants of risk, before refining models to include less significant contributions. Among their suggestions are to compare the effect on the risk estimate of using several predictive microbiology models appropriate to the problem. They further advised that if process variations have a more profound effect on the estimate of risk than the choice of predictive microbiology model does, it is most efficient to use the simplest model available.

Methods to assess the performance of predictive models are available and some models perform well for some foods under controlled conditions. In practice, however, in most situations the complexity of systems being modeled in combination with potential for exponential increases or decreases in pathogen dose from farm to fork means that risk assessment models based on currently available predictive microbiology models are unlikely to yield reliable estimates of *absolute* risk. At this time, models probably have greater utility in estimating the magnitude of the effect of different risk-affecting factors or evaluating the value of different options to manage or minimize risk.

While modeling efficiency is an important consideration in stochastic simulation models, it should also be remembered that the inception of formal risk assessment was not as an academic exercise but to establish microbiological safety regulations for foods in international trade that were based on full use of objective scientific evidence. This was needed to remove artificial impediments to international trade in food so as to facilitate economic advancement of developing nations. As such, it is also desirable to develop scientifically rigorous and defensible models and to ensure that important details that can profoundly influence the fate of pathogens in foods are not overlooked for the sake of simplicity. As emphasized in this chapter, there are situations in which the ecology of bacteria and fungi in foods can behave chaotically, with radical changes in response to relatively small changes in predictor variables. These situations are normally near to limits for growth, under conditions that lead to the induction of cell stress responses or that involve competition between species, such as the Jameson effect. This theme and examples are considered in greater detail below.

Interactions and Correlations: Microbial Ecology of Foods Made Harder

In stochastic modeling, distributions are used to represent variable quantities to generate the range and likelihood of possible outcomes from the overall process or system that is being analyzed. During the simulation, combinations of values are drawn more or less randomly from the different distributions. Values in different distributions can, however, be correlated and failure to recognize and model these correlations can lead to simulation of scenarios that could never occur in practice. An example, involving *B. cereus*, was alluded to earlier. Variability is known to occur in the rates of growth, lower temperature limits for growth, and rate of thermal inactivation of *B. cereus* spores and could be represented by distributions in a simulation model. However, it is also known that greater heat tolerance in strains is inversely correlated with the ability to grow at low temperatures. A risk assessment for *B. cereus* in dairy products would need to consider the ability of spores to survive thermal processing and for surviving cells to be able to germinate and grow at refrigeration temperatures. Failure to recognize the inverse correlation between thermal tolerance of spores and ability to grow at low temperatures could lead to simulation of unrealistic scenarios and overprediction of risk.

This section discusses some other examples of correlations that may need to be considered when assessing microbiological risk from foods.

Shelf life and storage conditions

To assess food-borne microbiological risk from perishable foods, consideration has to be given to the normal shelf life of the product, because time prior to consumption will often have a strong influence on the fate (and number) of pathogens that may be present. Many foods in which fungal or bacterial pathogens can grow will have a shelf life that is limited by the growth of microorganisms that cause spoilage of the product. The expected shelf life (e.g., "best before" date) of perishable foods is usually based on storage at ideal conditions. Recognizing that deviations will occur, food manufacturers usually specify a conservative "use by" or "best before" date that allows for less than ideal storage conditions.

Environmental factors that favor the growth of the pathogen of concern often will also favor the growth of the spoilage organism, and under inappropriate storage conditions foods may spoil before the expected shelf life, i.e., the shelf life of the product is reduced. Equally, better than expected storage conditions (which in most cases will mean better than anticipated refrigeration conditions) can also occur, leading to longer potential times between product manufacture (or the point of contamination) and its consumption. Thus, there is often an inverse correlation between shelf life achieved and storage conditions that lead to fast growth rates of pathogens. Failure to recognize this correlation in modeling could lead to overprediction of risk, in particular, given the exponential nature of microbial growth.

One approach to account for this phenomenon in risk assessment is to model the growth of spoilage organisms in parallel with the growth of the pathogen of concern. Model iterations in which spoilage precedes consumption of the foods can be either eliminated or considered to contribute no risk (or reduced risk if it is considered that spoiled product might still be consumed), so that unrealistic combinations of time and temperature (or other relevant storage conditions) are not modeled and included in the risk estimate. Such scenarios are likely to predict high pathogen loads and would have greater weight in estimates of risk, potentially skewing the results to higher risk estimates. This problem is compounded because predicted microbial loads usually change as an *exponential* function of time, but risk is proportional to the absolute number of pathogens.

Jameson effect

Related to the above is the common observation that when one species/group of microbes in an environment, such as a unit of food, reaches its maximum population density, other species in that environment stop growing as well, at whatever population density they had achieved to that time. This phenomenon has been described as the Jameson effect. It is like a race between organisms in which the race is halted as soon as the first organism "finishes" the race (i.e., achieves stationary phase) so that further growth of other organisms is suppressed. If a benign organism "wins," further pathogen growth is minimized and can result in a risk level many orders of magnitude lower than if the pathogen "wins." Conversely, in situations in which benign competing microorganisms are eliminated through processing, subsequent contamination by a pathogen can result in significantly higher levels of its growth and risk. The outcomes of the "race" are a potentially complex interplay of initial levels of both the pathogens and competing microbes, their lag times and their respective growth rates, in turn a function of the product formulation and storage conditions, and represent another example of a dynamic system the outcome of which is sensitive to initial conditions.

While the generality of the Jameson effect for all species has not been systematically demonstrated, it is particularly relevant to products in which lactic acid bacteria reach high levels but without immediately causing overt spoilage of the food, such as in vacuum-packed processed meats, fish, and so on. An example is the risk from *L. monocytogenes* in MAP-ham, where lactic acid bacteria can reduce risk, for which data are shown in Fig. 4. In these situations, the role of lactic acid bacteria has been termed "the silent fermentation" because lactic acid bacteria suppress the growth of other microbes but without causing the types of changes associated with fermented foods. (Note: in many fermented foods microbiological stability is achieved through the combined effects of acidity, reduced water activity, organic acids and other microbial metabolites, and the interest from a risk assessment or food safety perspective is the rate of *inactivation* of pathogenic contaminants.) In many other situations, at the levels of organisms required to induce the Jameson effect, the product will already be spoiled.

For lactic acid bacteria and *L. monocytogenes*, at least, the Jameson effect *does not* principally appear to require the production of inhibitory compounds that specifically inhibit other microbes; instead, it appears to be related to competition for nutrients, or perhaps mild acidification of the food environment and release of lactic acid. Similar to the need to model the growth of spoilage organisms on shelf life, discussed above, for some foods and microbial hazards, it may be necessary to consider the growth and cell densities of other bacteria because of their effect on the potential growth of pathogens.

Environmental influences on pathogen virulence

Virulence refers to the ability of an organism to cause disease and also to the measure of that ability. The virulence of food-borne bacterial pathogens can increase in response to stressful environmental conditions, and conversely, be reduced during periods of rapid growth. Thus, risk is affected not only by the effect of environment on cell numbers but also potentially on the virulence of the pathogen.

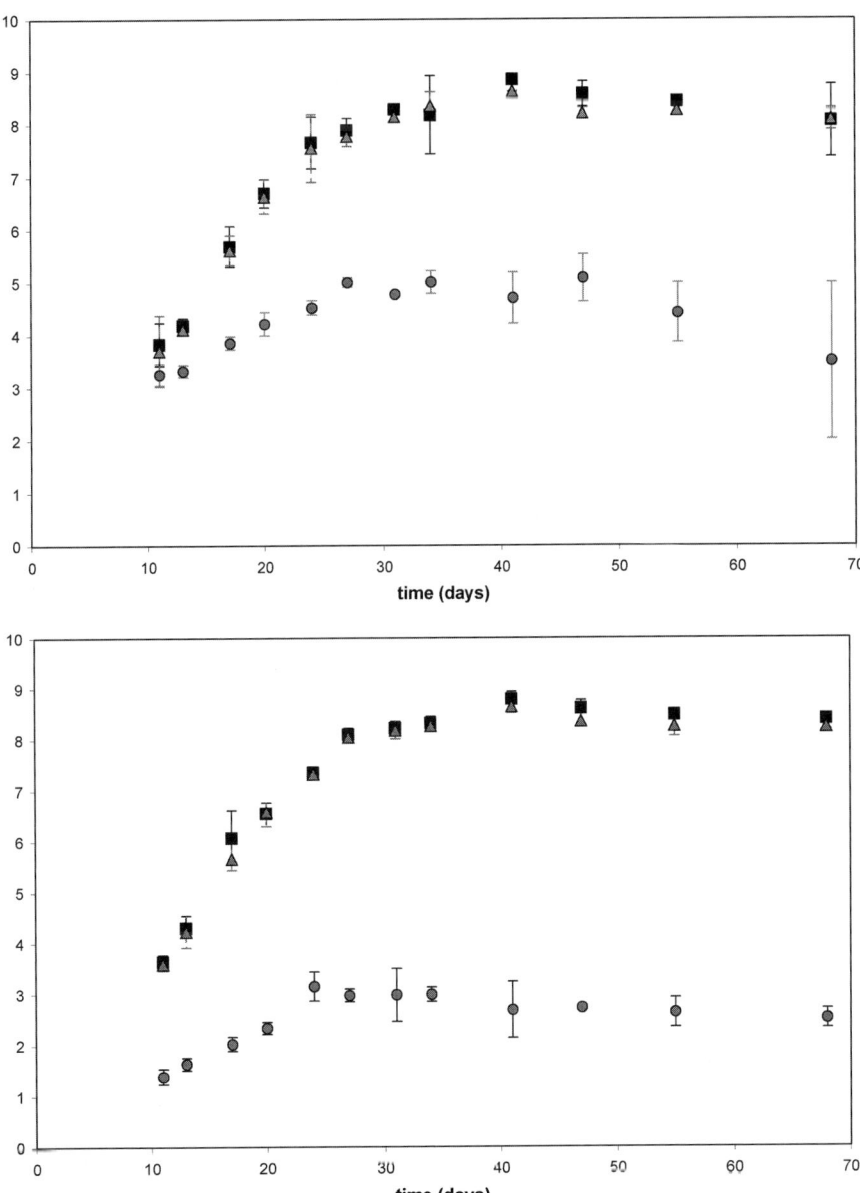

Figure 4 The coevolution of populations of total viable aerobic count (■), lactic acid bacteria (▲), and *L. monocytogenes* (●) in ham inoculated with *L. monocytogenes* and lactic acid bacteria. Error bars denote standard deviations of triplicate samples. In the upper figure, the level of *L. monocytogenes* is 100 times greater than that in the lower figure. Growth of the population of *L. monocytogenes* in either trial appears to cease when the total viable count achieves stationary phase. This is an illustration of the Jameson effect (F. Birrell, L. A. Mellefont, and T. Ross, unpublished data).

A typical bacterium contains several thousand genes that contain the information needed for production of proteins. "Expression" of a gene refers to the process of translating the sequence of nucleotide molecules contained in the gene (i.e., the "genetic code") into a protein that performs some function in the cell such as providing structure, environmental sensing, transport of other molecules, catalysis of chemical reactions, and so on. Proteins are also involved in virulence.

The virulence of a microbial pathogen can be considered to comprise two elements. The first is the ability to reach a site where it can establish an infection and the second is the set of biochemical characteristics that enable it to cause injury to the host, whether by physical damage to host cells, by the effects of host immune responses, or by production of toxins that interfere with the normal metabolism of the host. From this, two kinds of genes associated with the virulence of pathogens are recognized. The first, described as virulence genes, are those that encode proteins that are required for the cell to establish infection in the host. The second, described as virulence-associated genes, are those (i) that enhance the survival of the cell in hostile environments (e.g., conditions in foods and during food processing, or passage through the human stomach due to its acidity, exposure to bile in the duodenum and small intestine, etc.) and (ii) that improve the ability of the pathogen to spread to new hosts and overcome nonspecific defenses of the host.

While some proteins are required at all times by the cell, not all proteins are needed, or produced, at the same time. Specific molecules in the cell's environment can trigger expression of specific proteins, but in some situations the expression of whole suites of genes is coordinated and controlled by a single molecule. The changes are induced in response to environmental signals by moderation of cell molecules called "sigma factors."

Sigma (σ) factors are molecules produced by the cell that can control expression of *many* genes simultaneously. They do this by combining with a macromolecule called RNA polymerase that enables conversion of the genetic code into protein. A range of alternative sigma factors are known that are active in the cell under different sets of conditions. The suite of genes whose expression is governed by the activity of the sigma factor is the "regulon."

Under stressful conditions a particular type of sigma factor is produced by many species: in gram-negative bacterial cells it is the sigma-S (σ^S, also known as RpoS) and in gram-positive bacterial cells it is sigma-B (σ^B). σ^S was first described as an alternative sigma factor associated with responses in *E. coli* cells as they entered the stationary phase of growth. It is also present in *Salmonella*. RpoS mediates the expression of a large group of virulence-associated genes, including those involved in resistance to a variety of environmental stresses, such as starvation, hyperosmotic stress, oxidative stress (part of the immune system's response to invading cells), and low pH. Other sigma factors, such as Universal stress protein (UspA) and FliA, involved in synthesis and function of bacterial flagella (which confer motility to cells), are also considered to promote virulence-associated responses by their ability to resist environmental stress or through enhanced ability to reach sites where infection might be initiated. Motility is critical to the virulence of several bacterial pathogens including *Vibrio* spp. and *Yersinia enterocolitica*. Thus, stressful environments in foods can affect the ability of cells to survive, not only in foods and food-processing environments, but also in their passage through the human gastrointestinal system.

The alternative sigma factors induced by stress can also directly affect the expression of virulence genes. Among the bacterial food-borne pathogens there is strong evidence that expression of virulence factors in *L. monocytogenes* and *Salmonella enterica* serovar Typhimurium is increased in response to σ^B and σ^S, respectively.

Thus, stressful conditions not only increase the ability of food-borne bacterial pathogens to survive, they can also directly increase their ability to establish infection in humans. The role of stress and alternative sigma factors in enhanced virulence of other food-borne pathogens that have been studied, including *B. cereus*, *Staphylococcus aureus*, and enterohemorrhagic *E. coli*, is less clear with apparently contradictory effects observed by different researchers.

For more detail on this topic, Kazmierczak et al. (2005) provide a useful review, including a good introduction to the mechanisms of action and control of the activity of alternative sigma factors.

CONCLUSIONS: MODELING MICROBIAL ECOLOGY IN FOOD SAFETY RISK ASSESSMENTS

Fazil et al. (2005) observed that "many of the challenges faced by microbial risk assessors are related to issues such as dealing with rare events, the highly non-linear phenomenon of microbial growth and inactivation, the discrete nature of exposure to micro-organisms and the inherent complexity of the systems being described"—observations that are echoed by other workers. These issues relate both to mathematical needs such as the ability to accurately characterize the tails of distributions, as well as better microbiological knowledge. The preceding discussion has illustrated some of these difficulties and their genesis in the sensitivity of biological systems to initial conditions and to small changes in the environment.

One of the much proclaimed benefits the systematic approach of formal risk assessment offers to food safety management is the identification of important gaps in data and knowledge required to make better food safety decisions. In this regard, Fazil et al. continued more optimistically and added that "given sustained attention these and other technical problems are manageable and could be solved." Since the advent in the early 1990s of formal risk assessment approaches to management of microbial hazards in foods, that prediction has begun to be realized. Microbial food safety risk assessment is a strongly evolving field and one that is characterized by cooperation between microbiologists, modelers, statisticians, and food-processing experts. Publications addressing many of the issues identified in the preceding sections of this chapter have been presented and, given the short time since active research in microbial food safety risk assessment commenced, considerable progress has already been made. Areas in which numerous contributions have been made include:

- the effects of inoculum size/concentration on lag time, growth rate, and MPD;
- data and modeling approaches for pathogen transmission by cross-contamination, through the work of Schaffner's group in the United States and den Aantrekker and colleagues in The Netherlands;
- approaches for correct analysis of existing microbiological survey data (which is often based on presence/absence testing) for estimation of frequency of contamination and distributions of contamination levels in foods;
- modeling approaches for "competition" and other interactions between bacteria in foods;
- modeling methods for separation of uncertainty and variability in microbial risk assessments;
- approaches, including evaluation of their importance, for integration of stochastic modeling with fluctuating environmental conditions; and
- general methodological approaches including the Process Risk Model, the Modular Process Risk Model (MPRM) approach, strategies for selection of predictive models and appropriate implementation of model complexity, use, and evaluation of the potential advantages of Bayesian approaches, and so on.

The needs of risk assessment, in particular the characterization of variability of responses between strains, are being addressed by novel experimentation and new analyses of existing literature to provide better understanding and quantification of microbial physiology and ecology. New studies have examined the behavior, and variability, in responses of single cells and how this affects the responses observed in whole populations. Similarly, several studies have evaluated and reported variability in the growth behavior (e.g., growth limits, growth rates) of strains of the same species.

In published risk assessments, increasing model complexity and sophistication are evident over time, with more recent risk assessment differentiating and characterizing more sources of variability in the model inputs. This process has been termed "disaggregation," i.e., recognizing and modeling more variable inputs or more scenarios that lead to the risk. For example, some risk assessments have included the variation in pathogen MPD as a function of growth temperature; variability in the minimum growth temperature or lower pH limits for growth among strains of the pathogen of interest; use of a growth boundary model in conjunction with a growth rate model to determine whether growth is possible under the influence of combined hurdles to growth; recognition of different probabilities of pathogens being present in different phases/regions of cheese and the resultant differences in potential for growth; separate modeling of inactivation and subsequent growth of psychrotrophic and mesophilic strains of *B. cereus* to arrive at a single-risk estimate; cross-contamination of *Campylobacter* spp. between infected chickens and processing equipment and surfaces and vice versa; and so on.

However, disaggregation may mean that there are too few data to adequately characterize different inputs or that the increased number of variable inputs so broadens the output distribution that its value to support decisions is reduced. Thus, despite that additional input variables are recognized and could be included, sometimes it is more informative to group similar items together to better describe the variability in a larger, or aggregated, group (e.g., choose to "lump" all strains of a species together and describe them by an average response, and variability around the mean, rather than model each strain discretely). In other situations, modelers may choose to consider the effect of variation in one model input, while holding some others constant, to better resolve the influence of that input. Such simplifications can be useful and justified, in particular, if the inputs that are held constant are known to have minor influence. One of the main attributes of microbiological risk assessment, according to the guidelines developed by the Codex Alimentarius Commission (Codex, 1999), is that of transparency, requiring that, if simplifying assumptions are made, that they are clearly documented and, ideally, explained or justified. An example of an "acceptable" simplifying assumption relates to modeling temperature fluctuations in the farm-to-fork pathway. In practice, temperatures fluctuate, but in most risk assessments the temperature experienced by the microbe during a particular phase of the farm-to-fork pathway is assumed to be a constant value drawn from a distribution of possible values. Albert et al. (2005) showed that modeling microbial increases based on the mean temperature of a process that has fluctuating temperature did not produce a growth estimate that differed significantly from a model based on modeling the effects of the continually fluctuating temperature. This supports similar observations and calculations in Ross (1999).

One of the main lessons learned by those involved in development of microbial food safety risk assessments is the need to clearly define the question that the risk assessment is required to answer, or the decision for which additional knowledge or insight is required. With clear understanding of the aim of the risk assessment it is possible to limit its scope, make acceptable simplifications in the model, and expend more resources on modeling in detail the factors whose variation has the greatest influence on the risk. Thus, risk assessments that have a more narrowly defined focus tend to include greater detail.

From the discussion above, it is apparent that, despite the activity evident in the field, there is as yet no universal agreement on the best approach to modeling the microbial ecology of foods. In fact, the best approach may depend on the risk question. While there is general consensus that stochastic simulation models offer the best support for decisions, the main issues are to preserve the scientific and logical integrity of models developed and to clearly communicate the methods, data sources, and any underlying assumptions. This chapter has aimed to raise awareness of issues relating to the microbial ecology of foods that may be relevant to assessment of food-borne microbial risks. The issues highlighted are not intended to invalidate existing models. Rather, risk assessment is a collaborative and iterative process, and existing models can continue to be refined, improved, and expanded as more data, knowledge, and methodological approaches become available.

REFERENCES

Albert, I., R. Pouillot, and J.-B. Denis. 2005. Stochastically modeling *Listeria monocytogenes* growth in farm tank milk. *Risk Analysis* **25**:1171–1185.

Begot, C., I. Lebert, and A. Lebert. 1997. Variability of the response of 66 *Listeria monocytogenes* and *Listeria innocua* strains to different growth conditions. *Food Microbiol.* **14**:403–412.

Cerf, O. 1977. Tailing of survival curves of bacterial spores. *J. Appl. Bacteriol.* **42**:1–19.

Codex Alimentarius Commission (Codex). 1999. *Proposed Draft Principles and Guidelines for the Conduct of Microbiological Risk Assessment* (at step 5 of the Procedure). ALINORM, 99/13, Appendix IV, p. 58–64. (available from ftp://ftp.fao.org/es/esn/jemra/CAC_GL30.pdf).

Fazil, A., G. Paoli, A. M. Lammerding, V. Davidson, S. Hrudey, J. Isaac-Renton, and M. Griffiths. 2005. *Microbial Risk Assessment as a Foundation for Informed Decision-Making. A Needs, Gaps and Opportunities Assessment (NGOA) for Microbial Risk Assessment in Food and Water.* Microbial Food Safety Risk Assessment Unit, Public Health Agency of Canada, Guelph, Ontario, Canada.

International Commission on Microbiological Specifications for Foods (ICMSF). 1996. *Microorganisms in Foods 5: Microbiological Specifications of Food Pathogens.* Blackie Academic and Professional, London, England.

Kazmierczak, M. J., M. Wiedmann, and K. J. Boor. 2005. Alternative sigma factors and their roles in bacterial virulence. *Microbiol. Mol. Biol. Rev.* **69:**527–543.

McKellar, R. C., and X. Lu. 2004. Primary models, p 21–62. *In* R. C. McKellar and X. Lu (ed.), *Modelling Microbial Responses in Foods.* CRC Press, Boca Raton, FL.

McMeekin, T. A., R. E. Chandler, P. E. Doe, C. D. Garland, J. Olley, S. Putro, and D. A. Ratkowsky. 1987. Model for the combined effect of temperature and salt concentration/water activity on the growth rate of *Staphylococcus xylosus. J. Appl. Bacteriol* **62:**543–550.

McMeekin, T. A., J. Olley, T. Ross, and D. A. Ratkowsky. 1993. *Predictive Microbiology. Theory and Application.* Research Studies Press, Taunton, United Kingdom.

Mossel, D. A. A., and M. Ingram. 1955. The physiology of the microbial spoilage of foods. *J. Appl. Microbiol.* **18:**232–268.

Nauta, M. 2002. Modelling bacterial growth in quantitative microbial risk assessment. Is it possible? *Int. J. Food Microbiol.* **73:**197–304.

Roberts, T. A., and B. Jarvis. 1983. Predictive modelling of food safety with particular reference to *Clostridium botulinum* in model cured meat systems, p 85–95. *In* T. A. Roberts and F. A. Skinner (ed.), *Food Microbiology: Advances and Prospects.* Academic Press, New York, NY.

Ross, T. 1999. *Predictive Food Microbiology Models in the Meat Industry.* Meat and Livestock Australia, Sydney, Australia.

Ross, T., and P. Dalgaard. 2004. Secondary models, p. 63–152. *In* R. C. McKellar and X. Lu (ed.), *Modelling Microbial Responses in Foods.* CRC Press, Boca Raton, FL.

Ross, T., and T. A. McMeekin. 2003. Modeling microbial growth within food safety risk assessments. *Risk Analysis* **23:**179–197.

Sanchis, V. 2004. Environmental conditions affecting mycotoxins, p. 175–189. *In* N. Magan and D. Olsen (ed.), *Mycotoxins in Foods: Detection and Control.* Woodhead Publishing Ltd., Cambridge, England.

Shapira, R. 2004. Control of mycotoxins in storage and techniques for their decontamination, p. 190–223. *In* N. Magan and D. Olsen (ed.), *Mycotoxins in Foods: Detection and Control.* Woodhead Publishing Ltd., Cambridge, England.

van Gerwen, S. J. C., and M. H. Zwietering. 1998. Growth and inactivation models to be used in quantitative risk assessments. *J. Food Prot.* **61:**1541–1549.

Vose, D. 1996. *Quantitative Risk Analysis: A Guide to Monte Carlo Simulation Modelling.* John Wiley and Sons, New York, NY.

Wilson, P. D. G., T. F. Brocklehurst, S. Arino, D. Thuault, M. Jakobsen, M. Lange, J. Farkas, J. W. T. Wimpenny, and J. F. Van Impe. 2002. Modelling microbial growth in structured foods: towards a unified approach. *Int. J. Food Microbiol.* **72:**275–289.

World Health Organization and Food and Agriculture Organization of the United Nations (WHO/FAO). (2003). *Hazard Characterization for Pathogens in Food and Water: Guidelines.* Microbiological Risk Assessment Series 3. Food and Agriculture Organization of the United Nations, Rome, Italy.

World Health Organization and Food and Agriculture Organization of the United Nations (WHO/FAO). *Exposure Assessment of Microbiological Hazards in Food. Guidelines.* Microbiological Risk Assessment Series 7, in press. Food and Agriculture Organization of the United Nations, Rome, Italy.

Microbial Risk Analysis of Foods
Edited by Donald W. Schaffner
© 2008 ASM Press, Washington, D.C.

The Modular Process Risk Model (MPRM): a Structured Approach to Food Chain Exposure Assessment

4

Maarten J. Nauta

INTRODUCTION

The use of models to quantitatively describe the transmission of pathogens over the food-production chain is increasing in quantitative microbial risk assessment (QMRA) (Whiting and Buchanan, 1997; Cassin et al., 1998; Bemrah et al., 1998; Hartnett et al., 2001; Nauta, 2001; Nauta et al., 2001; Lindqvist et al., 2002; Rosenquist et al., 2003). Such models may cover the whole "farm-to-fork" food pathway or only the part that is relevant to the problem. They aim at assessing the consumer's exposure to the pathogen in food. Human risks can be evaluated by combining the exposure assessment with a dose-response relation. Because it can be used to evaluate the costs and benefits of proposed intervention measures (Havelaar et al., in press), food chain QMRA offers a valuable tool for food safety risk managers.

A general problem of QMRA is that the transmission of the pathogens through the food chains is often complex. First, there is a large variety of processes in the farm-to-fork food chain that require different models. The variety of models that can be used is large, and in a broader perspective a description of the food chain may be difficult. Next, for some parts of the food chain data may be abundant, whereas for other parts no appropriate data are available whatsoever. One is easily tempted to only model the data-rich parts of the chain and to jump over the other parts, which nonetheless may be equally important.

These considerations, probably recognizable for anyone involved in this type of risk assessment, stress the need for a structural, unified approach to

---

MAARTEN J. NAUTA, Laboratory for Zoonoses and Environmental Microbiology (LZO), Centre for Infectious Disease Control Netherlands, National Institute for Public Health and the Environment (RIVM), P.O. Box 1, 3720 BA Bilthoven, The Netherlands.

food chain exposure assessment. Here we explain the use of a methodology developed for this purpose: the Modular Process Risk Model (MPRM). This methodology builds further on the Process Risk Model developed by Cassin et al. (1998). It has been applied and developed in several food chain risk assessments since 2001, as illustrated below. This chapter aims to present and illustrate this methodology and to discuss how it may be used to simplify food chain QMRA in the future.

Food chain QMRA, as discussed in this chapter, is performed by the order of risk managers who are in close contact with the risk assessors. The primary aim of the risk management, in general, will not be to assess the risk, that is, a probability of an adverse health outcome due to microbiological contamination in the food chain, but to control it. Therefore the models are constructed as a tool to assess the effects of control measures, or in a broader context, "alternative scenarios." Such scenarios can represent all feasible changes in the processes along the food chain due to either risk management interventions or autonomous developments. As such, food chain QMRA is also a useful tool to identify crucial data gaps that have to be filled to properly assess the effects of control measures.

A QMRA can be set up either for industry or for the government, which may have different perspectives and different purposes and therefore may require different models. Governmental risk managers will be interested in food products and related risks on a (national) population scale, and in the context of (inter-) national food safety regulations. On the other hand, industrial risk managers will be predominantly interested in safety and risks associated with their own products and their own (well known) production processes. Examples of MPRM as discussed here do mainly deal with governmental food chain QMRA, but the methodology can be adapted by both.

MPRM offers a structured approach to exposure assessment as part of risk assessment. The end result is an assessment of the (distribution of) the number of pathogens consumed with the food by a population. It serves as an input for a dose-response relation, which is constructed in the hazard characterization stage of risk assessment. The risk is characterized when the exposure assessment and dose-response relation are combined. Discussion of the dose-response relation and risk characterization falls outside the scope of this chapter. It is assumed that a dose-response relationship, relating ingested doses with a probability of infection and/or illness, is available to complete the exposure assessment to a full-risk assessment. As a consequence, it is necessary to quantify the doses to assess health risks for individuals and populations. This implies that it is not sufficient to only know the frequency of people exposed to contaminated food products without knowing the numbers of microbes ingested or, in other words, to know only the prevalence of contaminated food products without knowledge about the concentrations.

Hence, food chain risk assessment requires quantitative modeling. Changes in numbers of pathogenic microbes along the food chain have to be described to be able to predict risks and the decrease in these risks that can be achieved by interventions in the food chain. As these numbers are variable and the dynamics are uncertain, complex (stochastic) models may have to be constructed. The major disadvantage of this necessity is that it is laborious and expensive. MPRM aims to offer a structure to reduce this disadvantage in a long-term perspective.

The food chain QMRA models discussed in this chapter are developed as simulation models of the food pathway, that is, the food chain including primary production, processing, transports, retail, food preparation, and consumer food handling ending with the ingestion of the food. The models describe the changes in levels of contamination of the food products with the microbial hazard concerned. The techniques applied in MPRM aim at performing these simulations in an efficient way, that is, including only those details that are considered relevant for the quantitative description of the microbial transmission over the food chain and the evaluation of the effects of alternative scenarios. In practice, the model of the food pathway is implemented in a computer program using Monte Carlo simulation, a technique widely used in quantitative risk assessment. This technique allows working with stochastic models and dealing with variability and uncertainty, which are important in risk assessment. However, the application of other techniques, not treated in this chapter, may be equally valid.

A Monte Carlo simulation program runs a large number of iterations. In principle, each iteration represents one potential true event of transmission along the food pathway, that is, the production process of one food product. A random sample is drawn from the probability distributions used as input for the model parameters. The model output is then again a probability distribution that allows us to evaluate "probabilities" of events and therefore to evaluate risks. Monte Carlo simulation has the advantage that it is widely applicable, comprehensible, and relatively easy to use. Important disadvantages of the Monte Carlo method are that it needs precise probability distributions for all input parameters, may require a very large number of iterations in complex models, is difficult to use when modeling (nonlinear) dependencies, and may not be an adequate tool when rare events have a major effect on the final result. If one of these disadvantages is dominant in the analysis, other techniques (like Bayesian analysis, fuzzy logic, and others) may be applied.

A discussion of modeling platforms and software packages falls outside the scope of this chapter. Here we first outline the MPRM methodology, offering guidelines to perform a food chain QMRA. Then some examples are given to illustrate these guidelines and to show how MPRM has been applied in practical situations. The last section discusses how MPRM can be used for further development of food chain QMRA in the future.

MPRM METHODOLOGY

The MPRM Food Chain QMRA Approach

The MPRM concept splits the food chain into modules. By first defining the modules, data collection and modeling are guided toward the efficient construction of a chain model. As shown schematically in Fig. 1, we propose an approach for the exposure assessment modeling in a food chain QMRA in which the risk assessors proceed as follows:

1. Define the statement of purpose

It is crucial to know the precise objective of the QMRA, that is, the question for which the risk managers seek an answer. Food chain modeling in general requires many simplifications and generalizations of the various processes, and these have to be made in the context of the statement of purpose. First, the microbial hazard(s) and the food product(s) considered have to be defined. Also, the population of consumers to which the risk assessment applies has to be stated.

Within risk analysis, the statement of purpose should be defined in agreement with the risk management. Before the model is constructed one should consider which alternative scenarios are to be evaluated. Such alternative scenarios are either interventions based on risk mitigation strategies or on other potential changes in the process that are a consequence of autonomous developments. The impact that interventions or autonomous developments may have on the risk will often be more important for the risk management than the risk estimate itself. Therefore, these scenarios are a necessary basis for further choices in the risk assessment process.

2. Give a description of the food pathway

It is essential to define the food production, handling, and preparation process that has to be modeled. For a single product manufactured by a single food manufacturer this may be straightforward, but if the risk assessment deals with a food product produced by many producers in one country or internationally (like broiler chicken meat products, milk products, etc.), it may be complex. Here either a set of food chains or a representative food chain has to be chosen for modeling, based on the statement of purpose. Processing

MPRM for Exposure Assessment 103

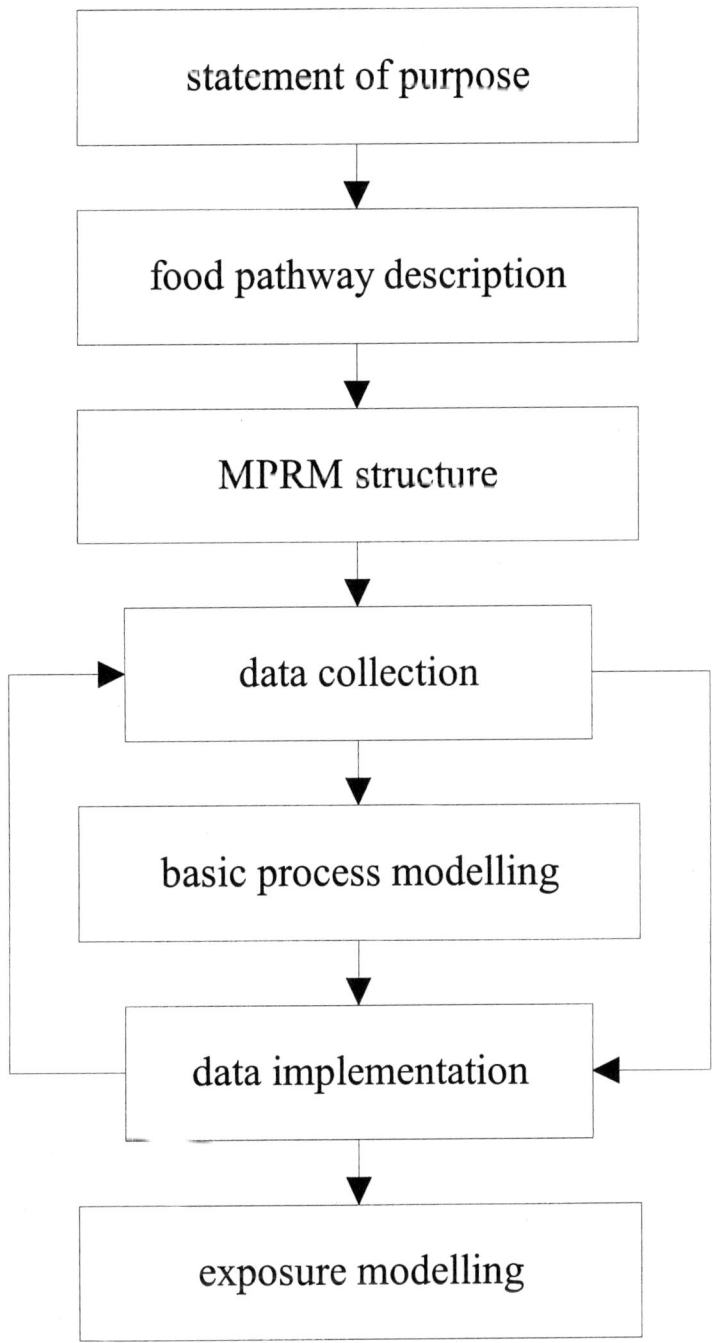

Figure 1 Seven steps for conducting QMRA with MPRM.

steps that involve potential alternative scenarios may need a more detailed description than processing steps that will remain unchanged.

3. Build the MPRM structure

The food pathway is then split into MPRM modules. In principle, each module is defined to reflect one of six basic processes: growth, inactivation, mixing, partitioning, removal, and cross-contamination, as explained in more detail below. A module will often cover a processing step as defined by the food processor, but this need not be the case. A module may combine several processing steps if they have a similar impact on the microorganisms. Also, if two or more basic processes interact in a processing step, one can decide to define a combined module. If a processing step is too complex or if essential parameters are unknown, it may be that the processing step cannot be assigned to any of the basic processes. In that case it can be considered as a "black box" process.

4. Collect the necessary data and expert opinion

Only at this stage should the search start for available data on the food pathway processes and the presence and dynamics of the hazard to be described in the MPRM modules. Gathering of data before this stage may imply the gathering of data that will later not be used in the QMRA or the gathering of data at an inappropriate level of detail.

Once the risk assessor has an overview of the available data, the next two steps in the MPRM approach can be taken. After that, it may be required to search additional data or to elicit expert opinion if appropriate data cannot be found. Here the use of expert judgment (preferably by a formal process [Cooke, 1991;Van der Fels-Klerx et al., 2005]) is preferred above "risk assessor judgment" to enhance the credibility of the risk assessment.

5. Define the model to use for each module

On the basis of the statement of purpose, food-production process and food-handling knowledge, data availability, and the alternative scenarios considered, the basic process models in the modules are defined and linked. The models used should preferably be based on the mechanisms of the process. This will give an increased insight in the dynamics of the process modeled and yield better model predictions when the input of the modules is modified, for example, as a consequence of management interventions. However, one should only put effort in the construction of complex models when this is necessary for evaluating alternative scenarios and when the availability of data permits it.

6. Implement the available data into the model

Available data are now precisely collected and then implemented in the models. Here it may appear necessary to find additional specific data and to reiterate the process, going back to step 4.

7. Perform an exposure assessment

By now the MPRM is finalized and an exposure assessment can be performed. Alternative scenarios can be analyzed by running the model with modified input or modified model parameter values. Also, uncertainty and sensitivity analysis of the model can be performed.

MODULES

To describe the transmission of a microbial hazard along the food pathway, the food pathway is represented as a chain of modules. As illustrated in Fig. 2, for each module we are interested in the changes in (i) the number of pathogens (N) per unit of food product, and (ii) the prevalence (P) of units of food product contaminated with the pathogen, that is, the fraction of units with $N > 0$. Note that, to ensure the correct implementation of the particulate nature of microbes, the number N is defined as an integer, not a concentration. Doing so, we use only whole numbers in our calculations, which prevents the possibility of accidentally using "fractions of individual cells" as living entities that can grow out or be infectious. As a consequence of this approach, in each module it is important to precisely define the "unit" of food product (e.g., a carcass, a bottle, or a package) and the "unit size" (weight, volume, or surface area). Units and unit sizes typically change along the food chain.

The aim of the model of each module is to describe the input-output relationship of N and P. The new prevalence (at the end of the basic process) is given by P_{out}, and the new number of microorganisms is given by N_{out}. A change in unit size is usually defined in the food pathway description. P_{out}

Figure 2 Each module i in MPRM describes the change in prevalence (P) and number of microorganisms per unit (N) in a process step. The food chain can be regarded as a line of linked modules.

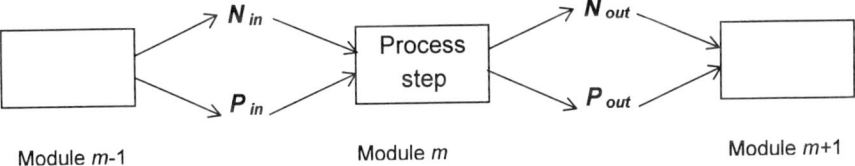

Module *m*-1   Module *m*   Module *m*+1

and N_{out} will be a function of N_{in}, P_{in}, and the unit size, and usually of some process parameters. P_{out} and N_{out} of a process step (i) will be the input P_{in} and N_{in} of the next process step ($i + 1$).

Essentially, this input-output relationship can be obtained either by observation (either direct in the production process [surveillance] or indirectly in an experimental simulation [challenge testing]) or by mathematical modeling (see, e.g., Notermans et al., 1998). Observational data have the advantage that it is process, product, and microorganism specific, but the disadvantage is that it is usually time consuming and expensive. Challenge testing is quick and cheap, but its disadvantage is that it may be imprecise and not specific for the process studied. The major advantage of modeling is that it is an instrument that allows an easy comparison of a broad range of alternative scenarios, the main objective of risk assessment. Therefore we focus on mathematical modeling here.

Variability and Uncertainty

Risk assessment deals with risk, which is a function of probability and severity of an event. Therefore risk assessment models calculate probabilities and use probability distributions that describe variability and uncertainty (Vose, 2000; Nauta, 2007). "Variability" represents a true heterogeneity of the population that is a consequence of the physical system and irreducible by additional measurements. One type of variability is stochasticity, where heterogeneity is a consequence of randomness, like the number of heads and tails after repeatedly tossing a coin. This can be considered essentially different from interindividual variability, which is a description of differences between members of a population, like the variability in height of children in a school class. "Uncertainty" represents the lack of perfect knowledge of the parameter value, which may be reduced by further measurements. In principle, this lack of knowledge can be quantified based on some assumptions and beliefs. The confidence interval that results from a statistical analysis, for example, usually serves as such a quantification of uncertainty. Other uncertainties may be very hard to quantify, especially if all data on a process step or a model parameter are missing.

As variability and uncertainty can both be represented by probability distributions and the difference between the two is not always obvious, they are easily mixed up. As shown by Nauta (2000), mixing of variability and uncertainty should be avoided in risk assessment, because it may lead to improper risk estimates. If variability and uncertainty are not separated, the probability distributions used to calculate risk at the end of the food chain can be a mixture of uncertainty and variability. This may give a biased insight in the risk estimate.

Some guidelines may be helpful in properly incorporating and separating uncertainty and variability in a process risk model.

1. In the simulation model, pretend that you know everything that can be known. You can do this by assuming that all the data you need are available. If all parameter values in a model are known, and the model properly describes the system to be modeled, there is no uncertainty. Everything in the model now described by probability distributions can be attributed to variability. If you then go back to the original model, with the data that are really available, you have also identified the uncertainty.

   This method can be applied rather easily in simulation models of food-production processes as described here. It is not easily applied in data analysis where it is not clear whether variability in results is to be interpreted as variability or uncertainty. However, when using these data in a risk assessment model, one can again pretend that it is clear how the data should be interpreted and analyze the model given this assumption. The result of this analysis can be compared with results for other assumptions, which then gives insight in the importance of separating uncertainty and variability at this point.

2. Make explicitly clear what the applied probability distributions describe. Variability can be variability in time, space, population(s), products, and so on. By explicitly identifying what the variability stands for, these variabilities will not be mixed up unnoticed. This method helps to prevent distributions describing several sources of variability in one, but it does not cancel out uncertainty.

3. A probability distribution describing variability can be uncertain if the variability is not exactly known. This is usually the case if insufficient data are available. The best way to describe such a "mixed" distribution is to first describe the variability (e.g., using the previous guideline) and then use new probability distributions to describe the uncertainty of the parameters describing the variability. So if, for example, the variability within a population is given by a Normal distribution with parameters μ and σ, the uncertainty can be described by probability distributions of μ and σ. This approach is known as second-order modeling (Vose, 2000).

4. Uncertainty is usually very difficult to quantify. In that case the options are to neglect uncertainty and make qualitative statements on the uncertainty in the conclusions, or to assess the uncertainty throughout the risk assessment for *all* uncertain parameters by expert judgment or otherwise. To assess the uncertainty in the risk estimate, incorporating

uncertainty for some uncertain parameters but neglecting it for others, in general, is not a meaningful strategy. It may, however, be useful if the purpose is to assess the relative decrease in uncertainty of the final risk estimate by additional research on one specific risk factor.

In MPRMs the number of cells per unit N and the prevalence P can be variable and uncertain throughout the model. As a result, we are able to assess the uncertainty and variability in the final exposure, and, with that, the uncertainty in the final risk estimate.

Within a module, the prevalence P can be uncertain, but usually it is not variable. If a (large) number of food products is modeled, the variability in numbers of (living) cells of the microbial hazard is given by a variability distribution of N. P is then defined as the fraction of units where $N > 0$, which is a property of this distribution, and is not variable itself. It can only be variable if, for example, different production lines are compared, or if seasonal variation is explicitly incorporated into the model. The main challenge of the models for the modules is to describe the change in variability distribution of N_{in} to a variability distribution of N_{out}. The uncertainty about N and P is an additional modeling level, above the variability model.

Basic Processes

An important element of the MPRM approach is to identify each module as one of the six basic processes, which are the backbone of every process risk model considered. These basic processes are the six fundamental events that may affect the transmission of any microbial hazard in any food process. There are two "microbial" basic processes, growth and inactivation, and four "food-handling" processes, mixing, partitioning, removal, and cross-contamination. The "microbial" processes strongly depend on the characteristics of the microbial hazard, as the effects of environmental conditions on growth and inactivation differ per species (and even per strain). Essentially, the effects of the "food-handling processes" are determined by the food-handling process characteristics only. Apart from aspects like "stickiness" and cell clumping, which affect the distribution over the food matrix and may be different for different microorganisms, food-handling processes have the same effect on any microbial hazard.

A variety of models can be applied for each basic process. The precise model applied will depend on the context of the risk assessment. Here, an overview of the basic processes is presented with reference to some models that can be applied for each of them. For some basic processes detailed information on modeling techniques that can be applied in risk assessment, as developed by the author, can be found in the literature (Nauta, 2001, 2002, 2005; Nauta et al., 2005).

Growth

Multiplication, or growth, is a typical characteristic of microorganisms and other living organisms. Prevention of growth of pathogens or spoilage bacteria is the objective of food preservation and common food safety practices like chilled storage. In MPRM terms growth gives an increase in N, the number of microorganisms per unit, and is therefore a particular risk-increasing process. Yet, it does not affect the prevalence. Due to microbial growth, products that are contaminated at a level below detection, and therefore are considered safe at a certain moment of time, may become unsafe later. As growth may occur after industrial food processing, this largely complicates setting microbiological food safety standards.

Microbial growth, as a variable, uncertain, and complicated process, is widely studied in predictive microbiology research. Here we introduce in brief the modeling of growth as a MPRM basic process.

In general, a growth model has the structure

$$\log (N_{out}) = \log (N_{in}) + f(.) \qquad (1)$$

with $f(.)$ an (increasing, positive) growth function. This growth function can have many shapes, which are discussed in the literature (e.g., McMeekin et al., 1993; Whiting, 1995; Van Gerwen and Zwietering, 1998). For example, for exponential growth $f(t) = \mu t$ (where t is time and μ is the specific growth rate), and when using the Gompertz equation $f(t) = a \exp[-\exp(b - c\,t)]$, with parameters a, b, and c.

In MPRM food chain risk assessment, the selection of the "best" model depends on the statement of purpose, process knowledge, and data availability. If, for example, the effect of a change in the temperature regime is considered, the model will need incorporation of "temperature" as a parameter. If not, it may be omitted.

When predictive growth models are applied in food chain risk assessment, it is important to realize that many sources of variability may interfere with the model predictions (see Nauta, 2002). Traditionally, predictive models are deterministic and aim to describe "best estimates." In contrast, risk assessment models are stochastic and aim to describe probabilities of undesired events like growth to large numbers. Therefore, next to stochastic variability, which, in particular, may have a large impact on the lag phase duration in small populations (Baranyi, 1998; Kutalik et al., 2005), variability between strains and variability in growth conditions, both in time and space, may also have to be incorporated in the models.

Inactivation

Microbial inactivation is the opposite of microbial growth. Inactivation is a frequently applied food safety and food preservation strategy and may be the consequence of various treatments applied in food production and preparation, like heating, lowering pH, drying, freezing, and so on. It is characterized by a decrease in the number of organisms per unit N. If this inactivation results in a decrease to zero living cells in one or more units, the prevalence P will decrease too.

The general formula for modeling inactivation is similar to equation (1)

$$\log (N_{out}) = \log (N_{in}) - g(.) \qquad (2)$$

where $g(.)$ is an increasing inactivation function. As for growth, many inactivation models are available (e.g., Van Gerwen and Zwietering, 1998; Xiong et al., 1999). The most frequently used inactivation process is heating, and the most frequently used inactivation model is the Bigelow model, in which the inactivation rate is a function of temperature. This is a linear function in time (t) and has the shape $g(t) = t/D_T$, where the D_T-value is the decimal reduction time at temperature T.

If inactivation is considered as a stochastic process, the stochastic variability in inactivation can be predicted (Nauta, 2001). For this purpose, define the expected value (E), calculated with the deterministic equation (2), as $E(N_{out})$. If the probability of survival of a single cell during the process step is now $E(N_{out})/N_{in}$, the number of survivors has a binomial ($N_{in}, E[N_{out}]/N_{in}$) distribution.

In contrast to the growth process, inactivation can have an effect on the prevalence. Once the number of cells in/on a unit drops to zero, the prevalence will decrease. The probability of this happening in a unit can be derived from the equation above as $[1 - E(N_{out})/N_{in}]^{Nin}$, which is approximately equal to $\exp(-E[N_{out}])$ when $E(N_{out})/N_{in}$ is small. Hence, the predicted prevalence after inactivation, if all units are equal in size, is

$$P_{out} = P_{in}\{1 - [1 - E(N_{out})/N_{in}]^{Nin}\} \qquad (3)$$

Partitioning

Partitioning occurs when a large unit is split into several small units, as shown schematically in Fig. 3. Examples of this process are an industrial batch of food distributed over consumer packages, milk from a tank distributed over bottles, or a carcass split into a number of trimmings. With the partitioning model we have to describe the reallocation of the N_{in} cells (or spores, particles, colony forming units [CFU], etc.) present in a large unit over n small units. Basically, the N_{in} cells are distributed into n portions $N_{out, i}$ (with i: 1 . . . n) and the challenge is to find the appropriate model to describe this distribution, which ultimately results in a distribution of N_{out} resulting from a distribution of N_{in}. As the number of units in the food chain increases

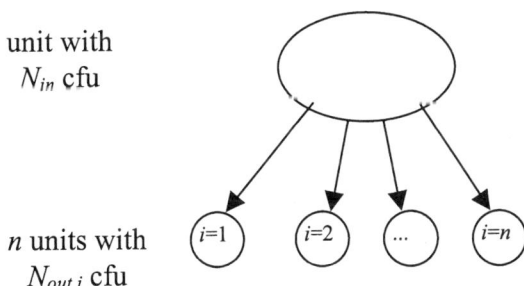

unit with
N_{in} cfu

n units with
$N_{out,i}$ cfu

Figure 3 Partitioning: A large unit, containing N_{in} cells (particles, spores, CFU, etc.), is split into n small units i ($i = 1 \ldots n$) that contain $N_{out,i}$ cells. The objective is to describe the distribution of the $N_{out,i}$ over the small units, given the values of N_{in} and n.

with partitioning and the total number of cells in the system remains equal, the prevalence is likely to decrease.

Several models for partitioning are described and discussed elsewhere (see Nauta [2005] for details). The simplest model assumes a homogeneous distribution of cells in a single large unit over the small units. It allows the use of a Poisson distribution to describe the variability between the numbers in small units. If the dependence between the numbers $N_{out,i}$ in the small units derived from one large unit is incorporated, a multinomial distribution can be applied to describe partitioning. And if, in addition, cell clustering and/or unequal sizes of the n small units are to be incorporated in the models, the process can be simulated by using a multivariate version of the betabinomial distribution.

Mixing

Mixing is the opposite of partitioning. In a "mixing" process units are gathered to form a new, large unit, as shown in Fig. 4. Examples are feces from several animals contaminating one carcass, carcass trimmings joined together in a ground beef batch, vegetables mixed for a batch of puree, milk from separate animals mixed in a milk tank, and so on.

If the numbers of cells on or in all small units are known, summation can be used to model the effect of mixing on the number of cells per unit: if k units are gathered, with unit i containing $N_{in,i}$ cells ($i = 1 \ldots k$), the larger

k units with
$N_{in,i}$ cfu

unit with
N_{out} cfu

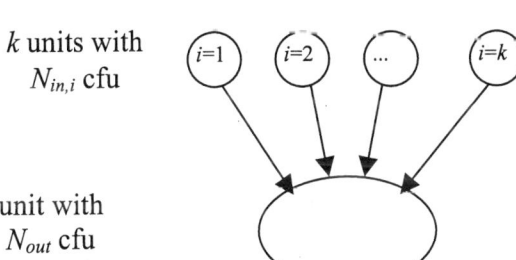

Figure 4 Mixing: k small units, containing $N_{in,i}$ cells (particles, spores, CFUs, etc.) in the units i ($i = 1 \ldots k$), are joined to form a new large unit with N_{out} cells. The objective is to describe the probability distribution of N_{out}, given a distribution of the $N_{in,i}$.

unit will contain the sum of all $N_{in,i}$ cells: $N_{out} = \Sigma_k N_{in,i}$ cells. Hence, the total number of cells in the system remains equal. With the unit size, the fraction of contaminated units (the prevalence) will increase. Assuming random homogeneous mixing and equally sized small units, the increased prevalence after mixing of k small units can be estimated as

$$P_{out} = 1 - (1 - P_{in})^k \qquad (4)$$

Complications related to the efficient and correct modeling of the mixing process and some practical solutions for food chain risk assessment modeling are discussed elsewhere (Nauta, 2005). In model simulations, mixing can be modeled by application of the Dirichlet distribution, a counterpart of the multinomial distribution as applied for partitioning.

An important issue is that application of the central limit theorem may not be appropriate to describe the effect of mixing as a summation process. This is particularly relevant if the variability between the numbers in a set of large units is considered in the risk assessment. The central limit theorem states that if $S_k = \Sigma_k X_i$ with X_i independent random variables with the same mean (μ) and standard deviation (σ), ($S_k - k\mu$)/$\sigma\sqrt{k}$ is asymptotically normal. In a mixing process this S_k stands for N_{out} and X_i stands for $N_{in,i}$. It suggests that, if k is large, the variability in N_{out} between several large units composed of k mixed small units from the same distribution can be described by a Normal ($k\mu$, $\sigma\sqrt{k}$) distribution, with μ and σ the mean and standard deviation of N_{in}. However, application of this distribution may be incorrect if the distribution of the $N_{in,i}$ is highly skewed, for example, when it is lognormal. In that case k has to be extremely large for a Normal distribution derived from the central limit theorem to apply. As a lognormal distribution is typical for microorganisms in many instances, and k will not be extremely large in most food-production processes, the central limit theorem cannot always be applied in process risk models (see also Nauta, 2001, 2005).

Removal

Several food-handling processes can be identified as removal. These fall into two categories. First, removal can be considered as a process where some units (or parts of units) are selectively removed from the production or food-handling process by the food manufacturer. Examples are the rejection of carcasses by veterinary inspectors in the slaughterhouse or the discarding of "ugly" vegetables. If this removal would be a random process with regard to the level of contamination with the hazard considered, it would have no significant effect on the risk assessment because only the food production will be affected, but not the distribution of numbers N. However, the process is usually performed because there is a presumed relation with microbial contamination: (heavily) contaminated units are discarded more often than lightly or noncontaminated units. Mechanistic modeling of this removal process needs information on the relation between the probability of rejection of a food unit and the number of hazardous microorganisms on food items. This sort of information is usually not available and in that case a simple model may be sufficient.

If it is assumed that contaminated units are removed with a higher probability than noncontaminated units, but heavily contaminated units are not discarded with a higher probability than lightly contaminated units, removal only affects prevalence, and not the variability distribution of the number of cells over the units. In that case, removal can be represented by a fraction f, such that the prevalence P_{out} after removal is equal to

$$P_{out} = P_{in} \cdot f / (1 - P_{in} + P_{in} f) \qquad (5)$$

where $0 \leq f \leq 1$. The rationale behind this equation is shown when equation (5) is rewritten as $P_{out}/P_{in} = f(1 - P_{out})/(1 - P_{in})$, or by expressing f as $f = (1 - p_c)/(1 - p_{not\,c})$, with p_c the probability of removal of a contaminated unit and $p_{not\,c}$ the probability of removal of a not contaminated unit. If $f = 0$, all contaminated units are removed; if $f = 1$, none of them are removed. This f may be variable and uncertain, derived from experimental results or expert opinion.

Selective removal of food units can also be applied as a risk management strategy when microbiological criteria are set in the food chain. This implies specific microbiological testing of the hazard concerned and may thus yield a good relationship between rejection and numbers of microorganisms N. This offers possibilities for testing and scheduling as discussed by Nauta and Havelaar (in press).

Another form of removal is washing, peeling, or an equivalent process. Here the removal process in principle aims at all units, yielding a similar model as an inactivation model, with a decrease in the number of cells N. So

$$\log (N_{out}) = \log (N_{in}) - h(.) \qquad (6)$$

where $h(.)$ is a positive "removal" function. If the removal process considered is a washing process (Nauta et al., in press), $h(.)$ can be assumed constant or variable for a process step. More complex models can be developed if the removal mechanistics are incorporated in the model.

Cross-contamination

Cross-contamination is an important process in food safety, but not well defined. Several types of cross-contamination can be considered.

Cross-contamination can be direct transmission of cells from one unit to another, e.g., by the (usually short and incidental) physical contact between two animals, carcasses, vegetables, and so on. It can also be indirect transmission, e.g., via the hands or equipment of a food processor. A third type is contamination from outside the specific food-production process, the introduction of cells from insects, dirty towels, and so on. This is usually referred to as recontamination, not cross-contamination. It may be an important process if it refers to the introduction of (substantial quantities of) the hazard considered into the food chain after the initial stage of the chain. If it is considered relevant, it should be incorporated in the exposure assessment as a separate source of contamination. An important issue is then whether this (re-) introduction is dependent on the first introduction as described by the food chain model, or not. If so, a relationship between these two has to be postulated. If not, the introduction can be modeled as an independent additional source of contamination, which is not discussed here.

The main concern about cross-contamination is that uncontaminated units get contaminated, thus increasing the prevalence. In principle, the total number of cells in the system remains constant as they are only redistributed over the units, but in practice cross-contamination will often coincide with a process like removal (Nauta et al., 2005) or partitioning (Nauta et al., in press), which may imply a reduction in the total number of cells involved.

For the change in prevalence a simple model can be applied, similar to equation (5) as given for removal, but now with $f > 1$ (Cassin et al., 1998):

$$P_{out} = P_{in}.f/(1 - P_{in} + P_{in}f) \qquad (7)$$

where $f > 1$. If $f = 1$, the prevalence is unaffected; otherwise, the impact of cross-contamination increases with f. The redistribution of N as a consequence of cross-contamination has been shown to be relevant in a risk assessment model for *Campylobacter* during poultry processing (Nauta et al., 2005). In that study, the cross-contamination model describes the reallocation of bacteria between carcasses and their environment in a production line. Because of this process the variability in the numbers of cells (N) per unit decreases, because heavily contaminated units lose cells to the lightly contaminated ones.

Here a novel simplified version of this model is presented. Consider a production line in which a series of units is processed in line through an environment to which cells are transmitted and back (see Fig. 5). This environment can be a transporting belt, a water tank, hands or equipment of a food handler, or anything by which cells can be transferred from one unit to a next one.

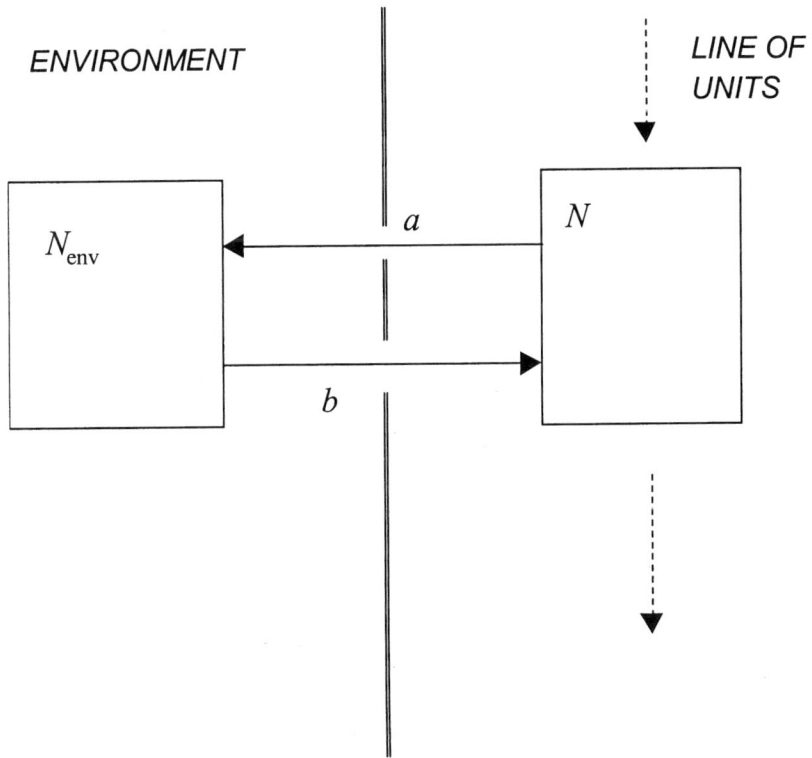

Figure 5 Cross-contamination between a series of units modeled in line (right) is assumed to occur via an environment that can represent air, water, equipment, hands, etc. The transmission rate to the environment is given by a, the transmission rate back is given by b.

The dynamics for unit i holding $N_{in,i}$ cells before the process and $N_{out,i}$ at the end, with $N_{env,i}$ in the environment after unit i passed the environment, can be written as a set of equations

$$N_{out,i} = (1-a)\,N_{in,i} + b\,N_{env,i-1}$$
$$N_{env,i} = a\,N_{in,i} + (1-b)\,N_{env,i-1} \qquad (8)$$

It can be derived that the expected value of $N_{env,i}$, $E(N_{env}) = a\,E(N_{in})/b$ (Nauta et al., 2005) and therefore, assuming the variability in the level of contamination in the environment is small,

$$N_{out,i} \approx (1-a)\,N_{in,i} + a\,E(N_{in}) \qquad (9)$$

This is not a surprising equation, as it indicates that (on average) the mean number of cells transferred *from* the units to the environment during cross-contamination [$a\,E(N_{in})$] is added *to* each unit again from the environment, after the environment reaches some kind of saturation. Also, with a fixed transfer rate a, $aN_{in,i}$ cells are transferred to the environment. As a consequence, the arithmetic mean of N is unaltered, but the variance decreases.

The effect of this type of cross-contamination on the distribution of N_{out} depends on the shape of the distribution of N_{in}, as illustrated in Fig. 6. If the distribution of N_{in} over the units is lognormal, the reduction for heavily contaminated units is negligible on a log scale, but the increase for the lowly contaminated ones on the same scale is large. As a consequence the geometric mean of the distribution of N increases. If the variability of N_{in} can be described by a Normal distribution, the decrease in variability affects both sides of the distribution, even on a log scale. Here, the geometric mean shows a very small increase.

These results are interesting for two reasons. First, in microbiology quantities of microorganisms are usually expressed in log units. It is shown here that the geometric mean, which is the mean of the logs, will increase due to cross-contamination of this type. Therefore means of count data on this process will show an increase, which is solely an effect of the decrease in variance, not of a true increase in the number of cells. Obviously, this may lead to a wrong interpretation of the data. Second, the right-hand tail of the distribution may be the dominant source of exposure and risk. It is shown (see Fig. 6) that the effect of cross-contamination on this right-hand tail depends on the shape of the distribution. Hence, the shape of the distribution describing the variability in concentration between food units may have a significant effect on the exposure and the risk. This strongly supports the need for quantitative methods in microbiological risk assessment.

Black box

Some processing steps in the food chain may have to be regarded as "black boxes." These processing steps are complex, and no data are available on the transmission dynamics of the hazard. It may include a combination of several basic processes as outlined above. If there is no proposal for an alternative scenario including this processing step, the transmission is most easily modeled by linear models on a log scale, that is, by assuming that the number of cells changes by an uncertain and possibly variable factor: $N_{out} = N_{in} x$, so

$$\log (N_{out}) = \log (N_{in}) + x, \quad (10)$$

with x a real number. The prevalence may change with a factor as modeled in the removal and cross-contamination basic processes

$$P_{out} = P_{in} \cdot f/(1 - P_{in} + P_{in} f) \quad (11)$$

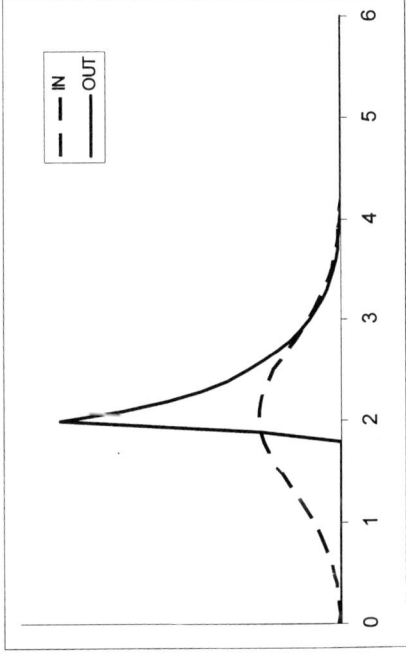
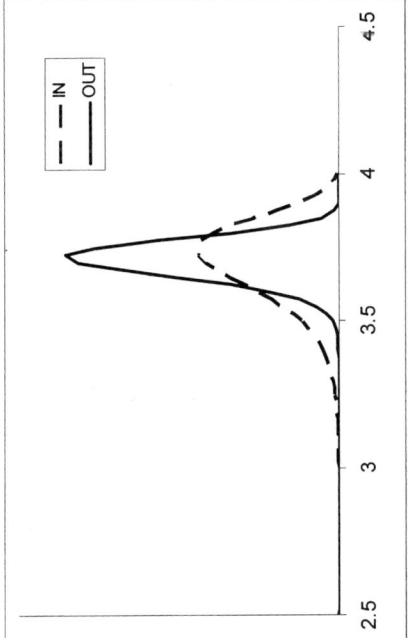

Figure 6 The effect of cross-contamination in a production line on the variability distribution of log(N), the number of cells per food unit. (Left) A normal input distribution of log(N_{in}), with $\mu = 2$, $\sigma = 0.7$. With model parameter $a = 0.2$ the output distribution strongly shifts to the right for low values but is almost unaltered for high values. (Right) A normal input distribution of N_{in}, with $\mu = 5,000$, $\sigma = 1,500$. With model parameter $a = 0.5$ the output distribution shows a smaller variance and also a decrease in high values.

where $f > 0$. The values of the factors x and f will have to be estimated by data from the process considered or expert opinion. In both cases variability and uncertainty will have to be incorporated explicitly in the model. If there are a series of input-output data available, these may show that the input-output relation is not linear. In that case another relation will have to be modeled, if possible, with a model based on the process mechanistics.

Black box models do not really fit in the MPRM methodology. The assumption of linearity between log N_{in} and log N_{out} is a crucial assumption, which may have serious consequences for the risk assessment if alternative scenarios are evaluated with modified input for the black box processing step, outside the range for which data are available. However, with restricted time and resources, black box modeling may be the only option, and should be emphatically indicated as such.

Input and Output of the Food Chain Model

If the MPRM, linking a series of modules describing the transmission of a microbial hazard, is established, the input and output of the QMRA food chain model require special attention.

The input of the model is in principle on the top of the food pathway. This can be at (the end of) primary production, but also at a later stage if that is sufficient for the objective of the risk assessment as formulated in the statement of purpose.

The model input should be a description of N_{in} and P_{in} for the first MPRM module, that is, a variability distribution of N_{in} between units and a value for the prevalence P_{in}. The uncertainty of both can be given by additional probability distributions.

Data on the prevalence are often available from monitoring and/or surveillance studies, but it is not always obvious how these data should be interpreted (Nauta et al., 2001). Next, data on concentrations of microorganisms are usually scarce. This means that the variability distribution for N_{in} for the first module can only be derived with quite some uncertainty. As we have shown in our risk assessment of *Campylobacter* in broiler chicken meat, this uncertainty can have a large impact on the final risk estimate (Nauta et al., in press). It is important to know not only the mean concentration, but also to characterize its distribution, because the final risk is largely determined by the (right-hand) tail of this distribution. The need for quantitative data on N_{in} is stressed in many food chain risk assessments, but unfortunately it is not easy to acquire those data.

The model output is an exposure assessment: a distribution of discrete doses (N_{out}) ingested by consumers eating the final unit in the MPRM. This distribution describes the variability in doses and will, of course, be uncertain. The prevalence P_{out} is the relative frequency of doses of one and higher. The distribution of N_{out} can serve as the input of a dose-response relation to calculate the mean probability of infection and/or illness in the population of concern. Here, the form of the dose-response relation should be such that it requires an actually ingested dose as a model input, and not the population mean dose (Haas, 2002). This implies, for example, that if a beta-Poisson distribution is known, one should apply a betabinomial distribution in a risk assessment that uses a Process Risk Model. Because the variability in the probability of infection between exposures is a quantity that need not and cannot be measured, it will rarely be of interest: The mean probability of illness of a population, which can be used to assess the human incidence related to the microbial hazard and the food product studied, is the final outcome of the food chain QMRA. It has no variability distribution, unless several populations (in time, place, consumption pattern, etc.) are compared.

EXPERIENCE WITH MPRM

MPRM was first applied in an exposure assessment of *Bacillus cereus* in a vegetable puree for a collaborative European research project (Nauta, 2001; Carlin et al., 2000; Nauta et al., 2003). Later we applied the methodology and developed it further in risk assessments on Shigatoxin-producing *Escherichia coli* O157 in steak tartare (Nauta et al., 2001; Havelaar et al., 2004) and thermophilic *Campylobacter* in broiler meat (Nauta et al., in press; Havelaar et al., in press).

Below we summarize two of these Dutch food chain risk assessments to illustrate how we implemented the MPRM food chain QMRA approach. Experiences that may be useful for risk assessors who want to apply MPRM are discussed. The summaries therefore focus on the approach and the methodology; their aim is not to discuss the risk assessments in detail.

B. cereus in Vegetable Puree

1. Statement of purpose

The aim of the food chain model is to perform an exposure assessment of *B. cereus* in 380-g broccoli puree packages produced by a European food company. End point of the exposure assessment is the prevalence of contaminated products and the number of colony-forming units in contaminated products at the moment that the consumer takes the puree packages out of the refrigerator for consumption.

As alternative scenarios, the effect of keeping to the "sell by date" that is printed on the broccoli puree packages and the effect of different temperature profiles in the refrigerators at the home of the consumer are to be analyzed.

2. Description of the food pathway

The food pathway was split in two, the industrial food-processing stage (see Fig. 7) and a retail and consumer stage. The latter consists of additional transport of packages to retail, storage during retail, transport home by the consumer, and storage at home. The description of the food pathway is based on information of the food company.

3. MPRM structure

The MPRM structure, in which the modules are defined, is based on the food pathway description and the characteristics of *B. cereus*. It is summarized in Table 1. A first analysis of the model showed that the model could be simplified because the cooking inactivates all vegetative cells and most of the *B. cereus* spores. Almost all *B. cereus* bacteria in the end product originate from the ingredients (milk powder and starch) added after cooking, so those on the raw vegetable could be neglected in the analysis.

4. Collect data and expert opinion

The available process data were unpublished data from the food company. Some missing data had to be based on expert opinion from either food company personnel or the European research team that collaborated in the project. Scientific literature data were available on growth and inactivation of psychrotrophic and mesophilic *B. cereus* strains, which differ in their ability to grow at low temperatures and in their heat sensitivity of the spores (Nauta, 2002; Nauta et al., 2003).

5. Define basic process models

The dominant basic processes were growth and inactivation, for which second-order predictive models from the literature were applied. In the food pathway, varying time-temperature profiles occur, but predictive models are not well suited to model these. Therefore some predictive models needed further development to meet our purpose.

6. Implement data in models

To cover the strain variety, a set of strains was selected for which both growth and inactivation data were available. Expert opinion was implemented in beta-PERT distributions (Vose, 2000) based on minimum, maximum, and most likely values. The whole model was constructed as a Monte Carlo spreadsheet model.

MPRM for Exposure Assessment 121

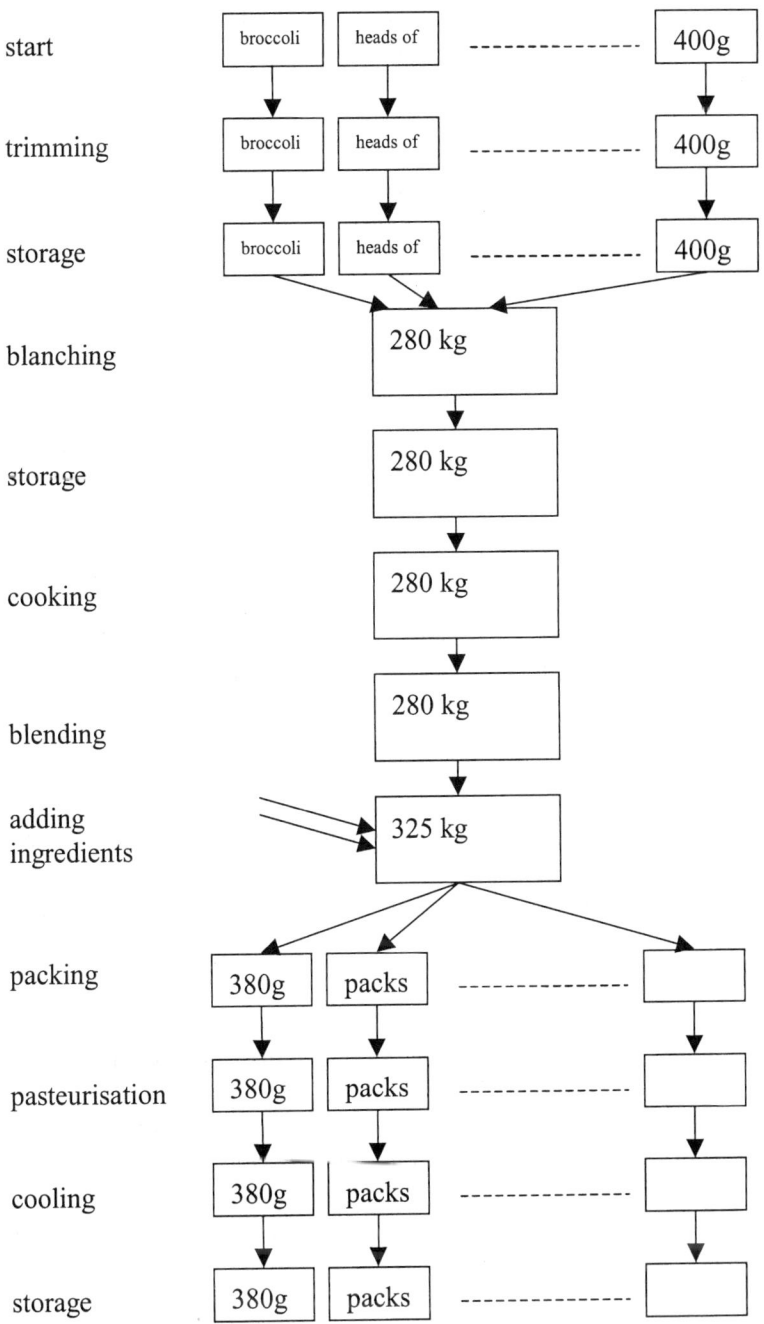

Figure 7 A schematic representation of the broccoli puree production process. The process starts with raw broccoli heads, which are treated and gathered in 280-kg batches. After ingredients are added, puree packages of 380 g are produced.

Table 1 The MPRM model structure of the food pathway as a series of basic processes[a]

Processing step		Basic process	Implemented basic process
1	Start		
2	Storage	Growth	
3	Trimming	Removal	
4	Storage	Growth	
5	Mixing and blanching	Mixing, inactivation	
6	Storage	Growth	
7	Cooking	Inactivation	
8	Blending	Inactivation	
9	Adding ingredients	Mixing, inactivation	START
10	Vacuum packing	Partitioning	Partitioning
11	Pasteurization	Inactivation	Inactivation
12	Cooling	Inactivation	Inactivation
		Growth	Growth
13	Storage	Growth	Growth
14	Transport	Growth	Growth
15	Retail	Growth	Growth
16	Transport	Growth	Growth
17	Home fridge	Growth	Growth

[a]As it appeared that cooking inactivates most *B. cereus* spores and the vast majority of the *B. cereus* in the puree packages originates from the ingredients, only the first stage of the production process is not incorporated in the analysis.

7. Exposure assessment

To evaluate alternative scenarios, the time for reaching critical levels of *B. cereus* was compared with the sell by date, and different distributions of temperatures in refrigerators were applied for different European regions, as based on published data. A summary of the results is illustrated in Fig. 8.

Lessons learned

- The recontamination step where ingredients were added was the main critical point in the production process.
- Better knowledge on the (variability in) storage conditions as applied by consumers was required for exposure assessment.
- High exposure was linked to a high probability of spoilage. With a simple method spoilage could be incorporated in the exposure assessment (Nauta et al., 2003).
- Variability between strains was important and had to be incorporated in the analysis. A distinction between psychrotrophic and mesophilic *B. cereus* strains was required because growth at low temperature had a large impact on exposure.

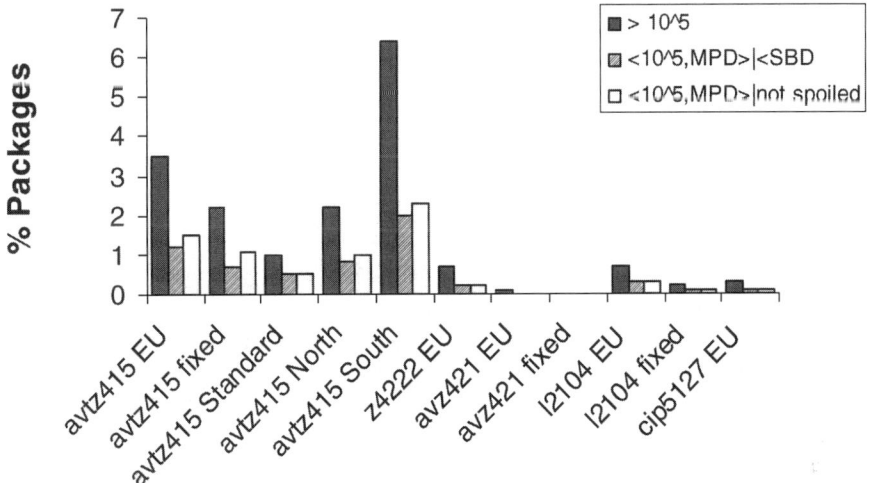

Figure 8 The percentages of packages containing critical levels of *B. cereus* for five different strain types (avtz415, z4222, avz421, l2104, and cip527) and a selection of five different "typical" consumer refrigerator temperature profiles (European [EU], fixed at 7°C [fixed], following regulations [= mean 5°C with standard deviation 1.9 {Standard}, North European {North} and South European {South}]). Percentages of packages with more than the critical level of 10^5 CFU/g are given as "> 10^5" bars. Other bars indicate more realistic values, leaving out packages that are likely to be spoiled with *B. cereus* cells after reaching the Maximum Population Density (MPD), together with either not passing the Sell By Date (SBD) and "not being spoiled" by spoilage flora, as predicted according to the time-temperature profile (Nauta et al., 2003). Apparently the psychrotrophic strain avtz415 with the "South" domestic fridge temperature profile (mean, 8.3°C) holds the largest risk.

- For spore-forming pathogens like *B. cereus*, sporulation and germination are important processes, but quantitative models that describe these processes are scarce.

Risk management advice

- By the lack of a dose-response model for *B. cereus*, risk assessment was not possible. (Based on the literature, exposure to levels >10^5 CFU was assumed to be critical.)
- For psychrotrophic strains the probability of a vegetable puree package containing more than the critical dose was larger than that for mesophilic strains. However, it has recently been found that this does not imply that psychrotrophic strains pose a higher risk (Pielaat et al., 2006).
- Decontamination of ingredients seemed to be the best control measure in this production process; another optional control measure was to improve the temperature control in consumer refrigerators.

Relevance of QMRA and MPRM

- The same analysis was performed using Bayesian Belief Networks, with identical results (Malakar et al., 2004). This confirms the validity of the methodology applied, without one being superior overall to the other.
- The quantitative analysis strengthens the risk management advice, which, however, might also have been derived by simpler methods. MPRM showed important data and knowledge gaps as mentioned above that are likely to be recurrent in other risk assessments as well and may have been overlooked otherwise.
- Models for inactivation of *B. cereus* are adopted and elaborated by others (Membré et al., 2006).

Campylobacter in Chicken Meat

1. Statement of purpose

The broiler meat risk assessment was part of the Campylobacter Risk Management and Assessment (CARMA) project in The Netherlands. The goal of this project was to advise the Dutch government about the effectiveness and efficiency of measures aimed at reducing campylobacteriosis in the Dutch population. Potential interventions in the broiler meat production chain were discussed with governmental risk managers and stakeholders from industry and consumer organizations.

2. Description of the food pathway

The food pathway is illustrated in Fig. 9.

3. MPRM structure

The line of basic processes is given as a column in Table 2. For chicken processing, it was not possible to effectively separate the inactivation, removal, and cross-contamination as basic processes. Growth of thermophilic *Campylobacter* is considered not relevant at the temperatures in the food pathway after slaughter.

4. Collect data and expert opinion

Some data were collected from literature, many model parameter estimates were obtained from expert opinion in a formal expert judgment study (Van der Fels-Klerx et al., 2005). In particular, the data on the variability of *Campylobacter* concentrations were scarce.

5. Define basic process models

A combined cross-contamination/inactivation/removal model was developed (Nauta et al., 2005). The inactivation process during consumer storage

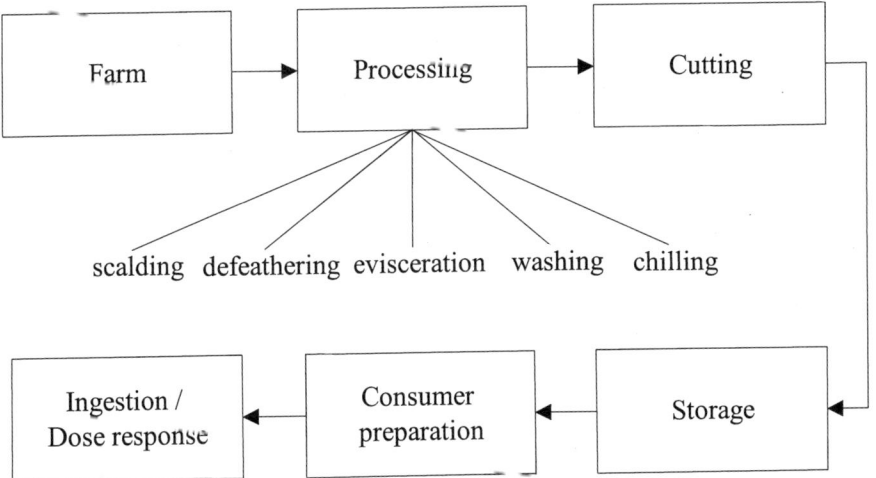

Figure 9 Food pathway of the *Campylobacter* risk assessment model. The MPRM exposure model started at the entrance of the processing plant and ended with salads cross-contaminated by fresh chicken breast fillet in a domestic setting.

Table 2 Overview of the models applied for each of stages of the risk assessment model[a]

Stage	Output[b]	Equation	Basic process	Unit
Scalding	$N_{ext, scald}$	See Nauta et al., 2005	c.c./rem/inact.	Carcass
Defeathering	$N_{ext, def}$	See Nauta et al., 2005	c.c./rem/inact.	Carcass
Evisceration	$N_{ext, evis}$	See Nauta et al., 2005	c.c./rem/inact.	Carcass
Washing	$N_{ext, wash}$	See Nauta et al., 2005	c.c./rem/inact.	Carcass
Chilling	$N_{ext, chill}$	See Nauta et al., 2005	c.c./rem/inact.	Carcass
Cutting (1)	$N_{ext, bc}$	$\sim \text{Betabinomial}(N_{ext, chill}(i), \alpha, \beta)$	Partitioning	Breast cap
Cutting (2)	$N_{ext, f}$	$= 0.5 \, (aN_{ext, bc}(i) + \varphi \, E(N_{ext, bc}))$	c.c./rem	(Single) filet
Storage	$N_{ext, fs}$	$= 10^{-r\_storage} N_{ext, f}(i)$	Inactivation	(Single) filet
Consumer preparation	N_{salad}	$= (t_{C,H} \times t_{H,H} \times t_{H,S}) + t_{C,B} \times t_{B,B} \times t_{B,S}) \, t_{S,S} \, N_{ext,fs}(i)$	c.c./rem	Salad (table companions)
Exposure	$N_{consumed}$	$\sim \text{Binomial}(N_{salad}, 1/v)$	Partitioning	Meal
Response	P_{ill}	$P_{ill}(n) = 1-(1-p_0)^{N\_consumed} \times P_{ill|inf}(N_{consumed})$		Meal

[a] Adapted from Nauta et al., 2005. See this report for an explanation of the symbols. c.c., Cross-contamination; rem, removal; inact., inactivation.
[b] The output of each stage is a function of the output of the previous stage.

describes the survival of *Campylobacter* at low temperature based on some experimental data.

6. Implement data in models

The expert data had to be transformed to model parameter values (Cooke et al., 2006). The whole model was constructed as a combination of two Monte Carlo spreadsheet models.

7. Perform exposure assessment

Results of the exposure assessment for salad cross-contaminated with *Campylobacter* from chicken breast fillet are given in Fig. 10. This figure also shows the results of risk characterization.

Lessons learned

- The model separated between *Campylobacter* in the chicken feces and on the bird/carcass exterior. This improved the processing model and allows evaluation of interventions aimed at campylobacters at different locations.
- Inclusion of variability in numbers of *Campylobacter* was crucial for the analysis.
- Variability was included in the quantitative model analysis, uncertainty was not. Although the uncertainty was included in the analysis of the expert judgment study, it was not possible to quantify the uncertainty at all stages of the chain model.
- The lack of quantitative data on initial contamination, and during processing, was evident. This is important because the inclusion of variability in numbers is crucial.
- The exposure model was linked with a dose-response model. The predicted incidence of human campylobacteriosis due to the consumption

Figure 10 The relative frequencies of dose classes (exposures) and the percentages of human cases of campylobacteriosis attributable to those classes. Most of the exposures are to low doses of 1 to 10 CFU *Campylobacter*. Higher doses have more impact than the distribution of exposures suggests.

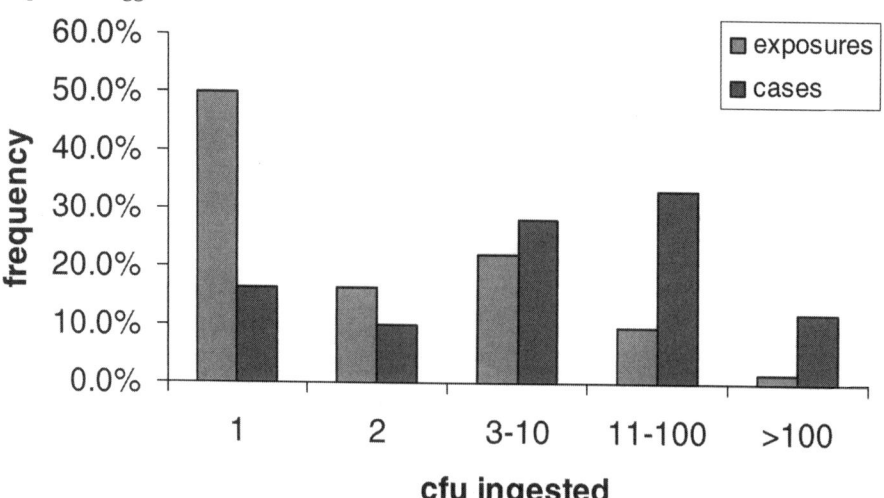

of salad cross-contaminated from chicken breast fillets was high compared with epidemiological estimates of the total incidence.
- However, compared with surveillance data, the prevalence of contaminated patties is underestimated by the model.
- The CARMA project included an economic analysis of interventions, a study of other routes of exposure, and a direct link with risk management (Havelaar et al., in press).

Risk management advice

- Management could be advised about the effectivity and the efficiency of interventions to control *Campylobacter* (Havelaar et al., submitted).
- In terms of risk reduction, decreasing concentrations of *Campylobacter* was least as effective as lowering the prevalence.
- Only irradiation of all fresh meat, or a ban on fresh meat, could guarantee fresh *Campylobacter*-free chicken meat.
- The risk assessment model can be developed further to link microbiological criteria in the food chain to Food Safety Objectives and Acceptable Levels of Protection (Nauta and Havelaar, in press).

Relevance of QMRA and MPRM

- Incorporation of variability between flocks and between birds/carcasses/fillets was essential for the risk assessment and the risk management advice. MPRM allowed doing this in a structural way.
- The relevance of incorporation of variability emphasized the need of a quantitative analysis and the need for quantitative microbial data.

DISCUSSION

MPRM and Other Risk Assessment Approaches

The MPRM so far has been applied predominantly at RIVM in The Netherlands. Initially, the approach built further on the Process Risk Model developed by Cassin et al. (1998). Other risk assessments (e.g., Bemrah et al., 1998; Rosenquist et al., 2003) apply similar methodologies but seem to lack a general framework. This illustrates that MPRM is not so much a novelty in the methods applied within the framework but mainly offers a transparent structure that may help tackling the complexity of quantitative food chain risk assessment. Its benefits are now recognized by other researchers as well (e.g., Lindqvist et al., 2002; Lammerding and McKellar, 2004).

One of the issues to be discussed before applying MPRM is whether a full quantitative analysis is needed. Quantitative food chain risk assessment is time consuming and therefore expensive, and it may be difficult to perform with a lack of quantitative data. Whether a quantitative approach is the best option to support risk management largely depends on the question of this risk management, as formulated, for example, in the statement of purpose of the risk assessment. Here one also may consider which part of the food chain is to be quantitatively modeled, as for example illustrated in the *B. cereus* risk assessment discussed above, where all steps prior to cooking could be ignored in the final analysis. For this purpose, a system for stepwise microbiological risk assessment allowing the main problems to be addressed before focusing on less important problems has been developed by Van Gerwen et al. (2000).

The need for a quantitative approach in risk assessment is illustrated in the *Campylobacter* risk assessment for CARMA (see above). Some conclusions of this risk assessment (like the negligible impact of scheduled processing to prevent cross-contamination between flocks and a ban on thinning at the farms) would not have been reached without a quantitative analysis. In this risk assessment it was evident that the final probability of exposure and the size of the ingested dose strongly correlate with the occurrence of high numbers of *Campylobacter* in some flocks higher up in the food chain. The (right-hand) tails of distributions have a large impact on the assessed risks. It is therefore required to include those distributions in the analysis and to describe them as well as possible. As illustrated in Fig. 6, mean and standard deviation alone may not give sufficient information for good risk assessment. This is even more important for lognormal (or comparable) distributions of concentrations of microorganisms, which are frequently appropriate: very high levels at one stage, followed by mixing or cross-contamination, may contaminate a large number of products (see above). In general, knowledge on or assessment of distributions describing the variability in concentrations or doses is required to properly assess exposure and the effects of alternative scenarios on exposure.

The basic models used in MPRM describe a process and therefore preferably have a mechanistic basis. This has the advantage that they incorporate nonlinear effects, based on what is known about the process considered. In contrast, models that are only derived from input and output data of a process are generally linear, potentially on a logarithmic scale. Such "data-driven" models do have the advantage that they are based on microbiological data but have the disadvantage that their validity to evaluate effects of alternative scenarios interfering with the process step modeled is obscure.

Some limitations of the MPRM methodology should be identified. First, it is a framework developed for postharvest quantitative food chain risk assessment. It is not suitable to model the dynamics of transmission and infection between living production animals, or the growth and survivals of pathogens in animal or human intestines. Here processes apply other than the basic MPRM processes. Second, it assumes that the food pathway can be simplified in a form of linked modules, representative for the system to be modeled. It may however be that, in a governmental food-commodity risk assessment, the variability in production processes of a food product is too large to do so. If production processes occur that are too complex for a mechanistic model, simplified "black box" models may be required. Third, the MPRM methodology as outlined here advocates the use of second-order Monte Carlo simulations, separating variability and uncertainty. As the examples above have shown, this separation of uncertainty and variability has been established, in part, on the basis of some assumptions about the interpretation of the data. However, we have not been able to quantitatively analyze the uncertainty in any case. An important reason for that is that we were not able to quantify the uncertainty in some of the modules of the food pathway and therefore could not characterize the total uncertainty either. Instead, uncertainty was explored by studying the impact of modified model parameter values on the exposure assessment results. Another reason is that the Monte Carlo simulation models were complex and time consuming with the incorporation of variability alone.

The use of MPRM also has some important advantages. The first is, of course, that it offers a clear structure to solving a complex problem. The definition of processing steps and modules cuts the food chain in comprehensible portions. The seven steps for conducting QMRA with MPRM are useful in preventing the risk assessor from starting to collect a wide variety of data that are available but are not useful for the risk assessment. By focusing on the statement of purpose, much (traditional) research on the problem at hand can be identified as irrelevant for the risk assessment, which reduces the amount of work to be done.

A second advantage of the MPRM methodology is that the availability of (basic process) models increases with the number of risk assessments done. A well-organized description of these models facilitates future risk assessments. Ideally, a library of MPRM basic models can be constructed for different pathogens and processes, which in the future may support food chain QMRA.

Using MPRM for Simpler Food Chain QMRA

As indicated, one of the major drawbacks of food chain QMRA is that it is time consuming and expensive. A food chain QMRA easily takes a few years' time, and at the end risk assessors still have a good reason to complain about a lack of research time. This conflicts with the needs of risk managers who quite often need quick advice on an emerging food-related health problem.

Hence, the request for quick risk assessments is increasing. One option for this is to simplify the risk assessment by not doing a (full) quantitative analysis. This indeed may speed up the process, but if the result is an insufficient risk assessment, the benefit of the quick result may be doubted.

Here it is proposed to use MPRM to simplify quantitative food chain QMRA. In fact, a simplified MPRM approach may be the first step after it has been decided that the identified problem requires more than a HACCP-like approach, possibly as one of the first steps in the stepwise risk assessment of Van Gerwen et al. (2000). In this approach, which is still in its infancy, the first three steps in the QMRA approach outlined above are maintained, yielding a series of modules and accompanying basic processes. Then the general effects of the basic processes are considered. As a first step, qualitative effects as given in Table 3 (Nauta, 2002) may give a quick overview of the potential impact of processing steps. Next, the general effects of the basic processes on the variability distribution of N, the number of cells per unit, can be studied. The latter may be particularly relevant if the right-hand tail of the distribution correlates with the final risk.

Figures 11 and 12 illustrate these general effects, assuming a lognormal distribution of numbers per unit N. This distribution is chosen because microorganisms often vary on a log scale and are also commonly presented as such. Note that the representations of the basic processes are simple versions of the model to meet the goal of simplicity.

Figure 11 shows that three basic processes result in lower levels of contamination per unit: inactivation, removal, and partitioning.

- Inactivation is modeled as a constant decrease in N, as in a deterministic primary inactivation model with a fixed time span, yielding a shift to the left of the entire distribution.

Table 3 Basic processes of the MPRM and their qualitative effect on the prevalence (P), the total number of organism in the system (i.e., all units evaluated in one simulation run of the model, N_{tot}), and the unit size[a]

Basic processes	Effect on P (the fraction of contaminated units)	Effect on N_{tot} (the total number of cells over all units)	Effect on unit size
Growth	=	+	=
Inactivation	−	−	=
Mixing	+	=	+
Partitioning	−	=	−
Removal	−	−	=
Cross-contamination	+	=	=

[a] =, no effect; +, an increase; −, a decrease. Depending on the precise definitions of the processes, removal may also give a decrease in unit size and cross-contamination may also give an increase in N_{tot}.

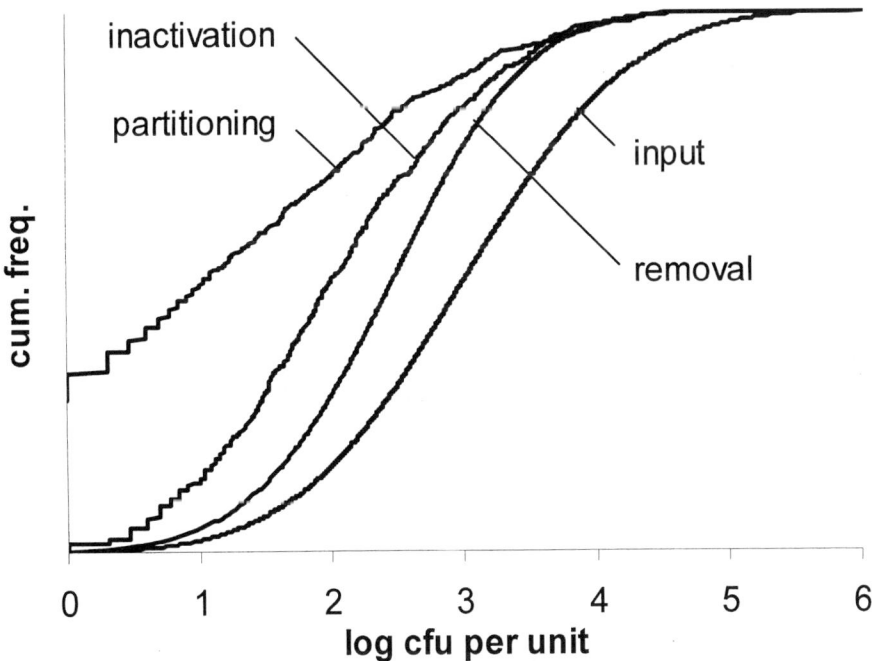

Figure 11 The effect of inactivation (1 log reduction), partitioning (to 10 smaller units), and removal on the log of the number of CFUs per unit, given a Normal N(3,1) variability distribution over the units of log N_{in}.

- Removal interpreted as a washing effect is identical with inactivation. Removal as a process that removes units with a higher level of contamination with a higher probability is modeled as a process in which units with a larger value of N are eliminated with a higher probability.
- Partitioning is modeled as nonhomogeneous partitioning to 10 equally sized smaller units, as a betabinomial process with parameter $b = 0.2$, indicating moderate clustering of cells (for details, see Nauta, 2005). The total number of cells in the system remains equal; the decrease in log CFU per unit is due to a decrease in unit size. Homogeneous partitioning yields a result almost identical with inactivation.

Figure 12 shows that the three other basic processes result in higher levels of contamination per unit.

- Growth is modeled as a constant increase in N, as in a deterministic primary growth model with fixed time span, yielding a shift to the right of the entire distribution.
- Cross-contamination is modeled by equation (9) with $a = 0.05$. N increases for the units on which it was low and slightly decreases on

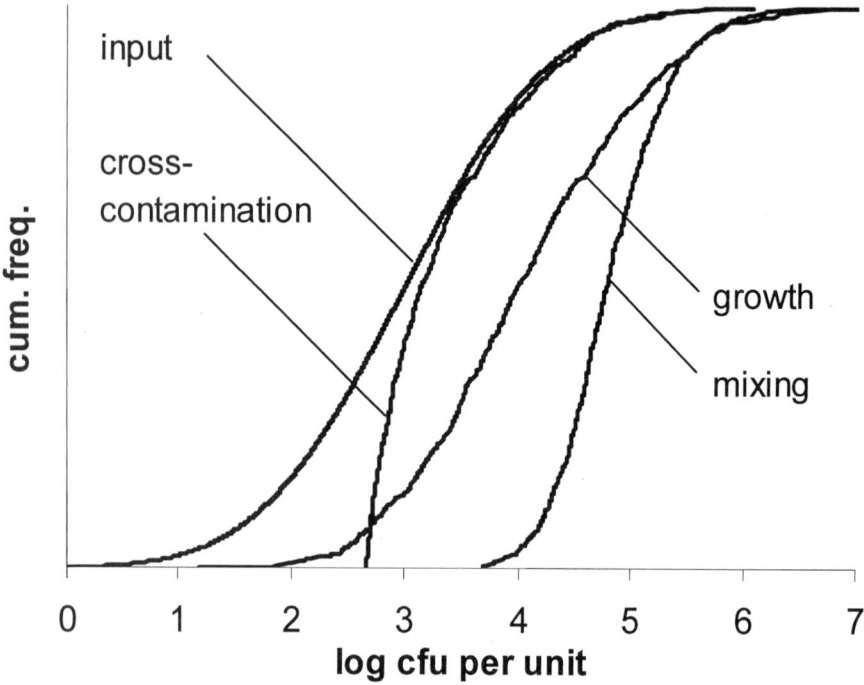

Figure 12 The effects of growth (1 log increase), mixing (of 10 units), and cross-contamination on the log of the number of CFUs per unit, given a Normal N(3,1) variability distribution over the units of log N_{in}.

those on which it was high. Due to the lognormal distribution of N, this decreasing effect for large N is negligible.
- Mixing is modeled as a homogeneous mixing process, in which 10 equally sized units form a new large unit. On average the new large units contain 10-fold the number of cells (an increase of one log), but here the randomly sampled units with large N result in a further shift to the right of the lower part of the distribution. As for partitioning, the total number of cells in the system remains equal.

In these results, the assumption of variability on a log scale is important, because the changes in shape largely depend on the fact that the arithmetic mean is much larger than the geometric mean (see also Nauta et al., 2005).

This comparison of effects of basic processes shows the following:

- It is not surprising that inactivation and removal interpreted as washing emerge as successful basic process to reduce numbers of cells: food producers frequently apply them to produce safe foods. However, note the effect of selective removal of units with a higher level of contamination.

Actually, it is the only process with a further shift to the left of the upper side of the curve. In practice "removal" can be performed in several ways, for example, by removing "ugly" food products, but also by doing microbiological tests on a set of units from the same batch. A good example of the latter is the practice of testing and scheduling of poultry flocks tested positive for *Campylobacter* or *Salmonella*. For the effectivity of such a removal process, the correlation between the selection criterion and the level of contamination with the pathogen considered is essential. Not only is the aim to remove the highly contaminated units, also one aims not to remove the uncontaminated units (Nauta and Havelaar, in press).
- Cross-contamination and mixing lower the variance in N between units, as shown by steeper curves in Fig. 12. Both processes will give an increase in the prevalence of contaminated units (not shown in Fig. 12). Both processes do not add microorganisms to the system but result in a reduced frequency of units contaminated with low levels. This will improve the effectiveness of the use of microbial testing of batches preceding a removal procedure as discussed above, because the correlation of levels of contamination between products and tested materials will be enhanced.

In future research it will be interesting to study the potential effects of shifting distributions like above on combinations of basic processes. For example, a combination of mixing and partitioning from and to the same number of units will result in a modified distribution with a lower variance. This may result in an (undesired) increase in the prevalence, but also in a (desired) reduction of heavily contaminated units, which may next enhance the effectiveness of a moderate inactivation step. A full analysis like this falls outside the scope of this chapter but may be a relevant exercise.

Considerations and studies like those mentioned above may help in simplifying quantitative risk assessment. By regarding the processes as identified in the MPRM structure as shifting effects on distributions, critical processes can be identified as in HACCP, but in a more quantitative way. These critical processes can then be studied with priority. Such an approach can probably also help in identifying the crucial data needs and identifying parts of the food chain where simple linear models are (in-)appropriate.

A last topic of further research may be the effect of shifting distributions on different end points of the risk assessments. Such end points may be the human incidence, exposure to a critical level, and also the relative risks of an intervention (Havelaar et al., in press). The (linearity of) the effects on these end points will probably differ, and improved insight in these effects may largely facilitate the potentials for simplifying food chain QMRA in the future.

REFERENCES

Baranyi, J. 1998. Comparison of stochastic and deterministic concepts of bacterial lag. *J. Theor. Biol.* **192**:403–408.

Bemrah, N., M. Sanaa, M. H. Cassin, M. W. Griffiths, and O. Cerf. 1998. Quantitative risk assessment of human listeriosis from consumption of soft cheese made from raw milk. *Prev. Vet. Med.* **37**:129–145.

Carlin, F., H. Girardin, M. W. Peck, S. Stringer, G. Barker, A. Martinez, P. Fernandez, W. M. Waites, S. Movahedi, F. Van Leusden, M. J. Nauta, R. Moezelaar, M. Del Torre, and S. Litman. 2000. A FAIR collaborative programme: research on factors allowing a risk assessment of spore-forming pathogenic bacteria in cooked chilled foods containing vegetables (FAIR CT97-3159). *Int. J. Food Microbiol.* **60**:117–135.

Cassin, M. H., A. M. Lammerding, E. C. D. Todd, W. Ross, and R. S. McColl. 1998. Quantitative risk assessment for Escherichia coli O157:H7 in ground beef hamburgers. *Int. J. Food Microbiol.* **41**:21–44.

Cooke, R. M. 1991. *Experts in Uncertainty: Opinion and Subjective Probability in Science.* Oxford University Press, New York, NY.

Cooke, R. M., M. Nauta, A. H. Havelaar, and I. Van der Fels. 2006. Probabilistic inversion for chicken processing lines. *Reliability Eng. Syst. Saf.* **91**:1364–1372.

Haas, C. N. 2002. Conditional dose-response relationships for microorganisms: development and application. *Risk Anal.* **22**:455–463.

Hartnett, E., L. Kelly, D. Newell, M. Wooldridge, and G. Gettinby. 2001. A quantitative risk assessment for the occurrence of *Campylobacter* in chickens at the point of slaughter. *Epidemiol. Infect.* **127**:195–206.

Havelaar, A. H., M. J. J. Mangen, A. A. De Koeijer, M.-J. Bogaardt, E. G. Evers, W. F. Jacobs-Reitsma, W. Van Pelt, J. A. Wagenaar, G. A. De Wit, H. Vander Zee, and M. J. Nauta. Effectiveness and efficiency of controlling *Campylobacter* on broiler chicken meat. *Risk Anal.*, in press.

Havelaar, A. H., M. J. Nauta, and J. T. Jansen. 2004. Fine-tuning food safety objectives and risk assessment. *Int. J. Food Microbiol.* **93**:11–29.

Kutalik, Z., M. Razaz, and J. Baranyi. 2005. Connection between stochastic and deterministic modelling of microbial growth. *J. Theor. Biol.* **232**:285–299.

Lammerding, A. M., and R. C. McKellar. 2004. Predictive microbiology in quantitative risk assessment, p. 263–284. *In* R. C. McKellar and X. Lu (ed.), *Modeling Microbial Responses in Food.* CRC Press, Boca Raton, FL.

Lindqvist, R., S. Sylven, and I. Vagsholm. 2002. Quantitative microbial risk assessment exemplified by Staphylococcus aureus in unripened cheese made from raw milk. *Int. J. Food Microbiol.* **78**:155–170.

Malakar, P. K., G. C. Barker, and M. W. Peck. 2004. Modeling the prevalence of Bacillus cereus spores during the production of a cooked chilled vegetable product. *J. Food Prot.* **67**:939–946.

McMeekin, T. A., J. N. Olley, T. Ross, and D. A. Ratkowsky. 1993. *Predictive Microbiology: Theory and Application.* John Wiley & Sons, Hoboken, NJ.

Membré, J. M., A. Amezquita, J. Bassett, P. Giavedoni, W. Blackburn Cde, and L. G. Gorris. 2006. A probabilistic modeling approach in thermal inactivation: estimation of postprocess *Bacillus cereus* spore prevalence and concentration *J. Food. Prot.* **69**:118–129.

Nauta, M., I. van der Fels-Klerx, and A. Havelaar. 2005. A poultry-processing model for quantitative microbiological risk assessment. *Risk Anal.* **25**:85–98.

Nauta, M. J. 2000. Separation of uncertainty and variability in quantitative microbial risk assessment models. *Int. J. Food Microbiol.* **57**:9–18.

Nauta, M. J. 2001. A modular process risk model structure for quantitative microbiological risk assessment and its application in an exposure assessment of *Bacillus cereus* in a REPFED. Report 149106 007. National Institute for Public Health and the Environment (RIVM), Bilthoven, The Netherlands.

Nauta, M. J. 2002. Modelling bacterial growth in quantitative microbiological risk assessment: is it possible? *Int. J. Food Microbiol.* **73**:297–304.

Nauta, M. J. 2005. Microbiological risk assessment models for partitioning and mixing during food handling. *Int. J. Food Microbiol.* **100**:311–322.

Nauta, M. J. 2007. Uncertainty and variability in predictive models of microorganisms in food, p. 44–46. *In* S. Brul, S. Van Gerwen, and M. Zwietering (ed.), *Modelling Microorganisms in Food.* Woodhead Publishing Ltd., Cambridge, United Kingdom.

Nauta, M. J., E. G. Evers, K. Takumi, and A. H. Havelaar. 2001. Risk assessment of Shiga toxin producing *Escherichia coli* O157 in steak tartare in the Netherlands. Report 257851 003. National Institute for Public Health and the Environment (RIVM), Bilthoven, The Netherlands.

Nauta, M. J., S. Litman, G. C. Barker, and F. Carlin. 2003. A retail and consumer phase model for exposure assessment of *Bacillus cereus*. *Int. J. Food Microbiol.* **83**:205–218.

Notermans, S., M. J. Nauta, J. Jansen, J. L. Jouve, and G. C. Mead. 1998. A risk assessment approach to evaluating food safety based on product surveillance. *Food Control* **9**:217–223.

Pielaat, A., L. M. Wijnands, K. Takumi, M. J. Nauta, and F. M. Van Leusden. 2006. The fate of *Bacillus cereus* in the gastro-intestinal tract. Report 250912005. National Institute for Public Health and the Environment (RIVM), Bilthoven, The Netherlands.

Rosenquist, H., N. L. Nielsen, H. M. Sommer, B. Norrung, and B. B. Christensen. 2003. Quantitative risk assessment of human campylobacteriosis associated with thermophilic Campylobacter species in chickens. *Int. J. Food Microbiol.* **83**:87–103.

Van der Fels-Klerx, H. J., R. M. Cooke, M. N. Nauta, L. H. Goossens, and A. H. Havelaar. 2005. A structured expert judgment study for a model of campylobacter transmission during broiler-chicken processing. *Risk Anal.* **25**:109–124.

Van Gerwen, S. J., M. C. Te Giffel, K. Van't Riet, R. R. Beumer, and M. H. Zwietering. 2000. Stepwise quantitative risk assessment as a tool for characterization of microbiological food safety. *J. Appl. Microbiol.* **88**:938–951.

Van Gerwen, S. J., and M. H. Zwietering. 1998. Growth and inactivation models to be used in quantitative risk assessments. *J. Food Prot.* **61**:1541–1549.

Vose, D. 2000. *Risk Analysis: A Quantitative Guide*, 2nd ed. John Wiley & Sons, Chichester, United Kingdom.

Whiting, R. C. 1995. Microbial modeling in foods. *Crit. Rev. Food Sci. Nutr.* 35:467–494.

Whiting, R. C., and R. L. Buchanan. 1997. Development of a quantitative risk assessment model for Salmonella enteritidis in pasteurized liquid eggs. *Int. J. Food Microbiol.* 36:111–125.

Xiong, R., G. Xie, A. E. Edmondson, and M. A. Sheard. 1999. A mathematical model for bacterial inactivation. *Int. J. Food Microbiol.* 46:45–55.

Microbial Risk Analysis of Foods
Edited by Donald W. Schaffner
© 2008 ASM Press, Washington, D.C.

Using Risk Analysis for Microbial Food Safety Regulatory Decision Making

5

Sherri B. Dennis, Janell Kause, Mary Losikoff, Daniel L. Engeljohn, and Robert L. Buchanan

INTRODUCTION

The use of risk analysis by federal regulatory agencies is a relatively new approach to respond to microbial food safety problems, even though the principles for risk analysis are increasingly being applied to food safety issues both domestically and internationally. Overall, risk analysis provides a structured framework to integrate knowledge about food-borne hazards, as well as the nature of the risks these hazards pose to consumers, combined with the technical and economic capacity to take appropriate interventions to significantly reduce food-borne disease. It allows the establishment of realistic, science-based targets to reduce the incidence of food-borne disease, plan and implement tailored interventions, and monitor the outcomes of these interventions. Thus, risk analysis provides the basis for establishing science-based standards integral to successfully control microbial food safety hazards. A risk analysis approach also emphasizes a shared responsibility for communication about food safety issues by all interested parties including government regulatory officials, industry, consumers, and academics. As a systematic process it both encourages participation and recognizes the unique roles and responsibilities of the various participating groups. As a science-based process it brings together science and policy in an effective manner to improve the safety of the food supply and human health.

SHERRI B. DENNIS, MARY LOSIKOFF, AND ROBERT L. BUCHANAN, Department of Health and Human Services, Food and Drug Administration, Center for Food Safety and Applied Nutrition, 5100 Paint Branch Parkway, College Park, MD 20740-3835. JANELL KAUSE, United States Department of Agriculture, Food Safety and Inspection Service, 1400 Independence Avenue, S.W., Room 333 Aerospace Center, Washington, DC 20250-3700. DANIEL L. ENGELJOHN, United States Department of Agriculture, Food Safety and Inspection Service, 1400 Independence Avenue, S.W., Room 349-E, Washington, DC 20250-3700.

137

Risk analysis is a process that includes risk management, risk assessment, and risk communication activities. The World Health Organization (WHO, 1995a) and the Codex Alimentarius Commission (CAC, 2001) have described risk analysis in terms of who is involved and when, what activities are conducted, and the sequence in which they are conducted. The three components of risk analysis are risk management, risk assessment, and risk communication.

Risk management is defined as the weighing of policy alternatives in consultation with all interested parties; considering risk assessments and other factors relevant to protecting consumers and promoting fair trade practices; and, if needed, selecting appropriate prevention and control options (CAC, 2001). Questions that risk managers might ask and answer upon being informed by the outcome of a risk assessment include the following: What is the problem (or risk)? How can we reduce the impact? Can we reduce the likelihood of the problem/risk? What are the options and trade-offs for addressing the problem/risk? What is the best way to deal with this problem/risk? What are the likely consequences of selecting a particular means of solving this problem?"

Risk assessment is defined as a scientifically based process of formally evaluating risks (CAC, 2001). A risk assessment typically consists of four main components: (i) hazard identification, (ii) hazard characterization, (iii) exposure assessment, and (iv) risk characterization. A risk assessment generally addresses questions posed by a risk manager. These questions may include: What can go wrong at one or more points in the farm-to-table continuum associated with the problem/risk? How likely is the problem/risk to occur? What are the consequences of the problem/risk? What factors can influence the problem/risk?

Risk communication is defined as the interactive exchange of information and opinions throughout the risk analysis process concerning hazards and risk; risk-related factors; and risk perceptions among risk assessors, risk managers, consumers, industry, the academic community, and other interested parties, including the explanation of risk assessment findings and the basis of risk management decisions (CAC, 2001). In conducting their work, risk communicators might ask both risk assessors and risk managers questions concerning: Who is the audience impacted by the problem/risk? What does the audience need to know? What do we need to hear from the audience? How can risk communicators best exchange information with the audience about the problem/risk? The risk communicators can also play an important role in ensuring that the risk managers and risk assessors are exchanging the information needed to effectively design and interpret risk assessments.

Hazards may be biological, chemical, or physical agents in food with the potential to cause an adverse health effect, whereas "risk" is a function of the probability of an adverse health effect and the severity of that effect, consequential to a hazard(s) in food. This chapter discusses the use of a risk analysis framework within a food safety context with emphasis on microbiological (biological) hazards, how risk analysis has been implemented in U.S. regulatory agencies and international organizations, and future needs to strengthen the use of risk analysis to help solve food safety issues. Case studies illustrate more specifically how microbial risk assessments have informed decision making in U.S. regulatory agencies.

WHY USE A RISK ANALYSIS FRAMEWORK FOR MICROBIAL FOOD SAFETY ISSUES?

Protection of public health remains a challenge in the United States and other countries. With continued globalization of food distribution, changes in consumer preferences, intensification of agriculture, urbanization, and new manufacturing and processing practices, food safety challenges are becoming more complex. The emergence of new food-borne pathogens, the reappearance of known pathogens in new foods, and the evolution of existing food-borne pathogens so that they are resistant to antimicrobial agents all add to the increasing complexity of food safety issues. Traditional food safety control measures, including good manufacturing practices, good agricultural practices, end-product testing and inspection, and consumer education have significantly improved the availability of safe foods over the past century but are limited in detecting and effectively dealing with the full range of evolving challenges posed by food-borne pathogens today. These traditional control measures have been limited primarily by the lack of insight regarding how interventions to control food-borne hazards correlated to a reduction in food-borne illness. As a result, too much emphasis was placed in some aspects of the food supply and not enough in others. An understanding of the entire food system from production to consumption was needed so that interventions could be identified and improvements effectively tailored to handle emerging and evolving food safety challenges.

The United States and other countries have taken steps to improve and strengthen their microbial food safety management systems by modifying the traditional approach of relying on prescriptive process restrictions and end-product regulatory inspections to one that focuses on a preventive science-based approach of controls throughout production, manufacturing, and marketing. A science-based approach considers available information about pathogen hazards associated with certain foods, their growth and decline under various conditions from production and processing through distribution and preparation, and epidemiological evidence linking the presence and amount of contamination in foods to food-borne illness. A milestone in the use of a science-based approach to food safety management was the adoption of Hazard Analysis and Critical Control Points (HACCP) system for foods produced in the United States and other countries. Although the HACCP system could encompass the entire farm-to-table continuum, in the United States, HACCP is most often used by slaughter, manufacturer/processors, and retail operations to (i) identify and evaluate the food safety hazards that affect the safety of their products, (ii) institute controls necessary to prevent those hazards from occurring or keeping them within acceptable limits, (iii) monitor and verify the performance of controls, and (iv) routinely maintain records to demonstrate that controls were operating properly. An integral part of HACCP is the establishment of critical limits that must be achieved at identified critical control points if a food system is to be considered "under control." In some instances, governmental agencies establish minimal performance standards that industry must meet to ensure that foods meet a specified level of safety (Hathaway, 1995). Establishing specific standards has been integral to the success of HACCP within the U.S. regulatory system; they define what will be considered safe by describing a verifiable minimal level of food safety control. Risk analysis, which can be used to better understand and integrate scientific data and information on the control of biological hazards in food as the food moves from production to consumption, could greatly enhance the effectiveness of establishing one HACCP performance standard over another to protect public health.

Overall, risk analysis provides a framework to integrate knowledge about food-borne hazards with the risks those hazards pose to consumers and the capacity to take appropriate interventions to significantly reduce food-borne disease. A risk analysis approach can provide the basis for:

- establishing science-based standards integral to the success of HACCP,
- allocating finite inspection resources among establishments that manufacture products posing the greatest risk to public health,
- evaluating effectiveness of different intervention scenarios,
- prioritizing of food safety research,
- developing specific food-handling information for consumers and retail establishments to further enhance the safety of foods,
- determining the equivalence of different approaches to controlling risks, and
- targeting messages to the most at-risk populations.

Benefits

Scientific research, surveillance data, and consumer surveys may be used to identify the factors that contribute to a public health problem and guide food safety decisions. Unlike the use of individual studies or scientific information, the value of using a risk analysis framework is in the ability to more effectively integrate a wide variety of information and data on hazards, foods, and human hosts to better understand their complex relationships and, more importantly, respond to food safety problems in a systematic and structured way. It allows the translation of disparate scientific data into useful public health strategies by conducting a risk assessment and translating that knowledge into scientifically defensible risk management actions. Risk assessment modeling, in particular, allows data on the presence and amount of microbial contamination in foods to be linked with a public health outcome (i.e., likelihood of illness and subsequent hospitalization and/or death). This capacity to link food safety activities with public health outcomes is leading to new concepts such as Food Safety Objective and Performance Objective. These new concepts are described in detail chapter 2 by Whiting and Buchanan.

In addition, coupling risk assessment with an economic impact analysis comparing risk management options may provide additional benefits: scientific assurance that selected food safety measures reduce food safety risks in a cost-effective manner, information on the magnitude of the benefit to public health expected by using one or more interventions, knowledge of the potential equivalence of various risk mitigation approaches, and assurance that the application of finite resources available to manage specific food safety risks is apportioned efficiently.

Moreover, the structured approach of risk analysis provides clarity with regard to the scientific basis guiding food safety management decisions and clear comparison of intervention options. With increased accessibility to risk analysis products (i.e., risk assessment reports and software models, and economic analyses available upon request and through federal and international websites), stakeholders have the information to better understand the food safety problem and possible solutions. The result is an improved public process and allows consideration of stakeholder input to guide federal and international food safety decisions (National Research Council [NRC], 1996).

Limitations

Despite all of the benefits, there are some limitations in using a risk analysis framework. A key limitation is the time required to conduct the risk assessment versus when the information is needed. In an emergency or crisis situation, sufficient time may not exist to construct a complex model. In some cases, the science is simple enough or the options are few enough, a risk assessment may not be a value-added endeavor, and a more simplistic (and possibly less certain) analysis or model would be sufficient.

It is possible for experts to quickly develop the first version of a model, which can then be followed by a more in-depth or complex model at a later date. For example, a model for *Enterobacter sakazakii* in powdered infant formula was developed by a panel of experts in less than 72 hours (Joint FAO/WHO Expert Meetings on Microbiological Risk Assessment [JEMRA], 2006). As more risk assessment models are developed and readily shared, future risk assessment models can be developed more quickly and modified to inform specific food safety decisions. For example, it took the United States several years to develop a quantitative risk assessment to evaluate the attribution of listeriosis among 23 categories of ready-to-eat (RTE) foods (Department of Health and Human Services, Food and Drug Administration [DHHS-FDA] / U.S. Department of Agriculture, Food Safety and Inspection Service [USDA-FSIS], 2001, 2003). The risk assessment was then used along with the development of a processing plant model to develop a second model in three months that identified those processing interventions that would be most effective in reducing listeriosis (USDA/FSIS, 2003a).

In considering the time frame required to conduct a risk assessment, it is important that sufficient time is allowed to adequately check the quality of the data used in risk assessment and that the models are sufficiently validated. Additional time is needed to allow for stakeholder participation and for peer review. In some instances targeted research may need to be commissioned to fill data gaps before the risk assessment can be completed or a decision made and this will add to the overall time frame. Comprehensive surveys and/or laboratory experiments may take months or years to plan, implement, and analyze the data, all of which adds to the time required to fully inform a decision. In addition, while stakeholder participation, peer review, and other review activities are critical and add value and insight to risk assessment, they also add to the time required to complete a risk assessment. For example, when a draft report is issued for public comment, the time frame established for stakeholder review of the report and submission of comments is usually in the range of 60 to 120 days. Once the public comment period ends, staff then need time to review the comments and make decisions on appropriate changes needed in the assessment and, if necessary, to rerun the model, revise, and finalize the report.

In addition, risk analysis requires adequate staffing and financial resources. The availability of expertise both within and outside an organization must be considered relative to the priority of a project and the specific staffing needs. If expertise must be obtained from outside the agency, financial limitations must be considered. In addition, financial resources are needed to fund research, data collection, and formal independent peer review of both influential and highly influential risk assessments to meet the information quality requirements established by the U.S. Office of Management and Budget (OMB, 2004).

IMPLEMENTATION OF A RISK ANALYSIS FRAMEWORK
International
Codex Alimentarius Commission (CAC) is an international body that has a long history of using risk assessment as a tool to evaluate pesticide and chemical contaminants. In about 1999, CAC began to explore the application of risk assessment to microbial food safety issues. With its role as the international food safety standards-setting body under the World Trade Organization (WTO) Sanitary and Phytosanitary (SPS) Agreement, CAC has accelerated its use of risk analysis for its many activities. CAC and its subject matter committees have developed or adapted procedures to apply the principles of risk analysis to highly complex and varied food control systems. In particular, the Codex Committee on Food Hygiene (CCFH) has encouraged the use of internationally focused microbial food safety risk assessments.

The CCFH provides advice on risk management and has proposed a framework for microbiological risk management consisting of four key steps: preliminary activities, evaluation of options, implementation, and monitoring/ review (CAC, 2005). (CAC defines a risk manager as a national or international governmental organization with responsibility for risk management. The focus is on governmental organizations with authority to decide on the acceptability of risk levels associated with food-borne hazards. CAC recognizes that this definition does not include all of the individuals involved in the implementation phase of risk management such as industry and consumers.) The preliminary activities include the identification of a microbiological food safety issue and the preparation of a risk profile document that describes the issue and its context, the current state of knowledge, and potential risk management options. If scientific advice (e.g., a risk assessment) is needed, the CCFH may seek assistance from the Food and Agricultural Organization of the United Nations and World Health Organization or other specialized scientific bodies (e.g., International Commission on Microbiological Specifications for Foods). Although an iterative process of communication should occur between the CCFH and the risk assessment group, in particular, early in the process to clarify the scope of the risk assessment, key assumptions, and the risk management options to consider, this process is particularly challenging since the CCFH only meets formally once a year.

The CAC has identified the following principles for effective microbial risk management (CAC, 2005):

1. Protection of human health is the primary objective.
2. It should take into account the whole food chain.
3. It should follow a structured approach.
4. The process should be transparent, consistent, and fully documented.
5. Risk managers should ensure effective consultations with relevant interested parties.
6. Risk managers should ensure effective interaction with risk assessors.
7. Risk managers should take account of risks resulting from regional differences in hazards in the food chain and regional differences in available risk management options.
8. Microbial risk management decisions should be subject to review and revision.

A regional organization involved in risk assessment is the European Food Safety Authority (EFSA), which was established in 2002 and serves as the primary body for food safety risk assessments for the European Union (www.efsa.eu.int). EFSA provides independent scientific advice on wide-ranging food safety issues including risk assessments related to both food and feed safety (including animal health and welfare and plant protection) and scientific advice on nutrition as it relates to European Community legislation. While EFSA's primary risk assessments are undertaken as requested from the European Commission, it also initiates its own work to address broader issues such as efforts to identify emerging food safety issues.

U.S. Federal Government

The USDA/FSIS and the DHHS/FDA have primary regulatory responsibility for ensuring the safety of the food supply (Dyckman, 2005). FDA is responsible for ensuring the safety and wholesomeness of domestic and imported food products (Federal Food, Drug, and Cosmetic Act [21 U.S.C 301 et seq.]), except meat, poultry, and egg products, which are the responsibility of FSIS (Federal Meat and Inspection Act [21 U.S.C 601 et seq.], Egg Products Inspection Act [21 U.S.C. 1031 et seq.], and Poultry Products Inspection Act [21 U.S.C. 451 et seq.]). Both agencies have increasingly relied on a risk analysis approach to guide national and international food safety decisions. Risk assessments have been used to provide scientific guidance for industry recall of potentially contaminated food, trade decisions, targeting of inspections and verification sampling resources, and, along with cost-benefit analyses, as required by OMB Executive Order

12866, guiding the establishment of cost-effective regulations that have a substantial (>$100 million) economic impact on industry. Both FSIS and FDA have been conducting quantitative food safety microbial risk assessments since the late 1990s, and follow the principles of risk analysis set forward by CAC. The specific procedures for conducting a quantitative microbial risk assessment are similar in both agencies.

WHAT MAKES A RISK ANALYSIS APPROACH WORK WELL?

The risk analysis process works well when each of its three components—risk assessment, risk management, and risk communication—is engaged and tied to the overall goal(s) to be achieved. Risk analysis requires the interaction of individuals with diverse skills and perspectives, including expertise in communication, coordination, project management, scientific knowledge, and skills in data analysis and modeling.

There are various ways to structure risk analysis within an organization. One approach is to use long-standing "dedicated" teams that are permanently staffed and another is to use "virtual" teams that are convened as needed. Each approach has merit. A benefit of having dedicated staffs is that it allows for the assignment of clear roles and responsibilities for providing organizational support to operate in a risk analysis framework. Having a permanent dedicated staff within an organization fosters an esprit de corps among risk analysts, building a common professional identity among a diversified group of technical experts with competing interests. For virtual teams, the benefits center primarily on assembling a team to accomplish a short-term project that is narrowly defined. This approach provides more flexibility to build a team composed of the specific subject matter experts needed to work on a specific issue. In general, this team does not have competing interests, such as the conduct of other risk assessments. In any organization, specific procedures for implementing the principles of risk analysis should be established taking into consideration what works well with the agency's culture, its legal authority, and the need for various types of risk assessments.

When using a team approach, the specific roles and responsibilities for each aspect of the entire risk analysis process must be clearly articulated. The responsible parties for various activities are summarized in Table 1 and discussed below. The risk management team is responsible for formulating the risk assessment questions to be addressed, providing key assumptions for the risk assessment, reviewing the draft risk assessment, developing risk management options to be assessed, and selecting and implementing the risk management policy decision. The risk assessment team is responsible for conducting the risk assessment and refining as necessary the assumptions

Table 1 Risk analysis activities and responsible parties[a]

Risk analysis activities	Responsible party		
	Risk assessment	Risk management	Risk communication
Select the risk assessment		√	
Plan and allocate resources	√	√	√
Performance			
Conduct the risk assessment	√		
Develop communication options		√	
Develop communication messages			√
Review			
Risk assessment documents	√	√	√
Evaluate and select management option(s)		√	
Communication messages	√	√	√
Issue			
Risk assessment documents	√		
Risk management action plan (policy decision)		√	
Risk communications messages			√

[a]Adapted from CFSAN, 2002.

provided by the risk management team, explaining the uncertainty of the results and the impact of assumptions on the results. The risk communication team is responsible for providing input to the risk assessment and risk management teams based on the identification and understanding of stakeholder concerns, information needs, and perceptions; promoting an ongoing exchange of information about the project with all interested parties; and developing public health messages based on the risk assessment results and risk management implementation plans.

Within the risk analysis teams, there are also specific roles and responsibilities. A team leader is typically identified and is responsible for the technical aspects of the work as well as for ensuring adherence to established timelines for completing the project, and in general serves as the spokesperson for interpreting or communicating about the risk assessment design. Depending on the scope and complexity of the work, the team leader may require the assistance of a project manager to assist with administrative and technical management of the process. It may also be helpful to have an individual assigned in the role of a science adviser who is responsible for ensuring that the science of the risk assessment is not compromised by the policy needs of the risk management team (Center for Food Safety and Applied Nutrition [CFSAN], 2002). The science adviser must act to preserve credibility and transparency in all decisions and is responsible for resolving any science issues.

APPROACHES TO DECISION MAKING

How do risk managers make decisions? A typical approach to solving a particular identified problem is to first gather information and data about the problem or issue. To do this, risk managers may commission a risk assessment, a literature review, new research, or consultation with experts both within and outside their organization. Next, the options for solving the problem may be identified and the strengths/limitations for each described and perhaps considered in relation to a cost-benefit analysis. Finally, a selection is made among the available options and the decision is implemented.

However, different risk management strategies may be needed based on the complexity of the issue, the level of uncertainty associated with the data and knowledge, and the ambiguity of the interpretation of the data. Renn (2005) offers five types of risk management strategies based on the characteristics of the risk issue:

- Simple. Routine or mundane risks that can be managed by using traditional approaches.
- Risk informed. Complex or sophisticated risk issues that can be informed by probabilistic risk assessment modeling and expert consensus.
- Precaution or resilience based. Highly uncertain risks that require balancing risks and benefits.
- Discourse based. Highly controversial risk with a high degree of ambiguity requires risk trade-offs and deliberation with stakeholders.
- Imminent dangers or crisis. Need for fast responses.

The discipline of decision analysis provides structured tools and skills to assist decision makers with their tasks (Clemen, 1996). This field of expertise is helpful, in particular, when uncertainties are associated with the various alternatives available to the risk manager.

Approaches to science-based decision making include the *technocratic*, *decisionist*, and *transparent* models (European Commission [EC], 2004). The *technocratic* approach assumes that policy decisions should be made solely based on scientific considerations, independent of social political, cultural, and economic conditions. In this approach, risk communication is in one direction, from government to stakeholders (e.g., industry and general public). Difficulties may arise in defending decisions made using this approach when large uncertainties exist in the science used as the basis for the policy decision. In the *decisionist* approach, science is considered first (for example, by conducting a risk assessment) and other factors, such as social, political, cultural, and economic, are considered independently (by risk managers) in formulating the policy decision. However, the *transparent* approach includes reciprocal links between science and policy. With this approach, first a risk assessment plan is developed to deal with any socioeconomic and political considerations. Next, the science is considered within the context of the risk assessment, and last, risk managers consider the technical data as well as economic and social information in making a regulatory decision.

USING RISK ASSESSMENT IN REGULATORY DECISION MAKING

The various activities that lead to a risk management decision and its implementation are illustrated in Figs. 1 and 2. Figure 1 provides a diagram showing seven components of risk management (trigger/input, prioritization, process, decision, implementation, outcome, and monitor/evaluate/modify) and the iterative nature of these components. Within the framework illustrated in Fig. 1, the conduct of a risk assessment is considered part of the "process" component, as one of many means of gathering data and information needed about a problem. The diagram shown in Fig. 2 provides a detailed flow chart to more specifically identify the activities associated with the conduct of a quantitative microbial risk assessment within a risk analysis framework. Both Fig. 1 and 2 illustrate many of the same types of risk management activities (although in some cases use different descriptive terms for these activities). For example, "triggers/inputs" is similar to "identify food safety problem," "prioritization" is similar to "risk analysis agenda," and "decisions" is similar to "select management options." Some of the key activities important to using a risk analysis approach for decision making are discussed below.

Identification and Prioritization of the Food Safety Problem

To efficiently allocate resources to address food safety issues, it is vital to establish annual food safety risk management priorities. National food safety priorities may be (i) triggered in response to food-borne outbreaks, (ii) driven by media attention and public concern, (iii) identified through the discovery of new scientific findings, and/or (iv) continued strategic focus to reduce food-borne illness to achieve a food safety goal. Prioritization is essential to allow a pragmatic evaluation of food safety challenges in light of competitive public health concerns (e.g., cancer, injury, obesity) and of the likelihood that there will never be sufficient resources to simultaneously address all food safety management issues. Improved epidemiological data on the attribution of food-borne illness to specific foods could provide much needed information to better prioritize federal efforts to reduce food-borne illness (Batz et al., 2004; Food Chemical News, 2004; Batz et al., 2005).

In addition, U.S. regulatory agencies must also ensure that they have the legal authority to act, because in the United States multiple agencies have

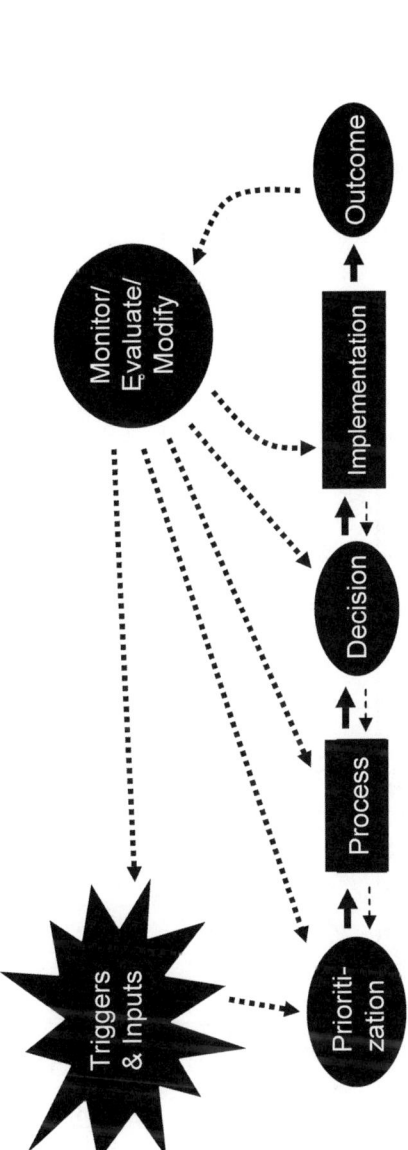

Component	Definition
Triggers/Inputs:	Triggers and inputs are indicators of food and cosmetics practice, public health or other issue(s) and concerns that may require a risk management decision. The results of a risk assessment, indicating a new need, may also trigger the framework.
Prioritization:	Weighing the relative importance, the capacity to accomplish and the timeframe for action. The identification and selection of risk assessments is described in FDA-CFSAN (2002).
Process:	The process is the means to gather and communicate data and information from both internal and external resources. A risk assessment is one means of gathering information. The procedures for conducting and managing a risk assessment are described in FDA-CFSAN (2002).
Decision:	Identification and selection of option(s) to mitigate, reduce or eliminate the risk in a practical, cost effective manner.
Implementation:	Acting on the decision; mobilize program resources through established or new procedures/guidance/policy/testing/investigation.
Monitor/Evaluate/Modify:	Active measurement and assessment of the outcome and determination if changes are needed.

Figure 1 A risk management framework. Source: Dennis et al., 2006.

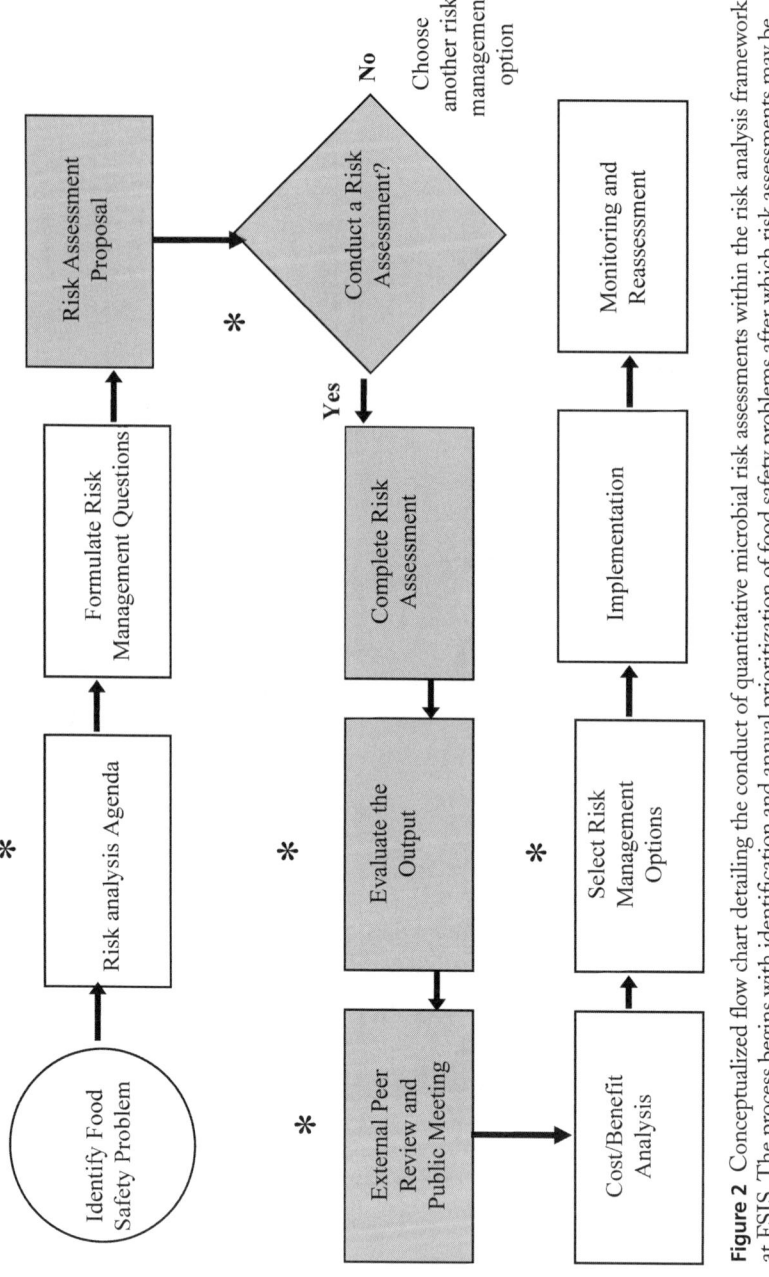

Figure 2 Conceptualized flow chart detailing the conduct of quantitative microbial risk assessments within the risk analysis framework at FSIS. The process begins with identification and annual prioritization of food safety problems after which risk assessments may be initiated to inform decision making by risk managers. The chart depicts risk assessment (shaded boxes) as part of a larger risk analysis framework. Asterisks (*) indicate those points at which input is solicited from stakeholders and others of the public. Adapted from USDA/FSIS, 2003b.

primary jurisdiction over the safety of animals and food production from farm to table and each may serve as independent risk managers.

Deciding Whether To Conduct a Risk Assessment

Note that a risk assessment may not be needed by decision makers when the problem is simple, it has only one defined solution, or the need for articulating the impact of the solution is not necessary because it is obvious (Buchanan et al., 2004). Factors to consider before commissioning a risk assessment are the time constraints for conducting the risk assessment, the available versus needed resources, the needed staff expertise and availability, and the availability of data and modeling tools. Once a decision is made to conduct a risk assessment, other key activities include developing a risk assessment plan, formulating risk management questions, and reviewing the risk assessment.

Risk assessment plan

A risk assessment plan or charge to the team should be developed that describes the type of risk assessment to be developed in light of specific risk management questions. The plan should include the scope of the risk assessment, including a description of the pathogen, food(s), at-risk populations, and end point of concern (illness, death) or model outputs (risk per serving, risk per annum). This plan should also describe the design of the risk assessment (i.e., conceptual model), the types of data and information to be used, how the uncertainty of the estimates will be handled (e.g., second-order modeling or sensitivity analysis), and key assumptions.

Establishment of a risk assessment plan is the responsibility of the risk managers requesting the risk assessment, but experience has emphasized that this is a process for which the risk assessors may need to assist the managers. It often takes substantial experience to be able to frame risk management questions in a manner that can be translated into inquiries that can be investigated in a risk assessment. The risk assessors should also be prepared to articulate in their risk assessments the impact that the risk plan, developed by the risk managers, has had on the risk assessment results.

The importance of a well-articulated risk assessment plan was highlighted in the report, "Science in trade disputes related to potential risks: comparative case studies" (EC, 2004). The report concluded that international disagreements over the legitimacy of food safety regulatory measures may occur because of implicit, unacknowledged, and unexamined assumptions. It recommended greater emphasis on identifying and clarifying all assumptions prior to commissioning a risk assessment, and thereby reducing the frequency of disputes about resulting policy judgments.

Risk management questions

Risk assessments should be designed to answer specific questions posed by the risk managers. A few examples of these questions from completed quantitative microbial risk assessments are provided in Table 2. Risk management questions initiate from the potential or intended use of a risk assessment. For example, the *"Listeria monocytogenes* in RTE foods" (DHHS-FDA/USDA-FSIS, 2003) risk assessment was used to establish priorities for surveillance and research and to identify new risk assessments needs. Risk assessments used toward the development of regulations include the *"Salmonella enteritidis* in shell eggs and *Salmonella* spp. in egg products" (USDA/FSIS, 2004a) and the *"Listeria monocytogenes* in deli meats" (USDA/FSIS, 2003a). The "Fluoroquinolone-resistant *Campylobacter* in chicken risk assessment" (FDA/CVM, 2001) was used to support the withdrawal of an approved product, and the "*Vibrio parahaemolyticus* in raw oysters risk assessment" (FDA, 2005) was used in the development of new regulatory initiatives.

Review

The quality of a risk assessment document, a risk assessment model, and the underlying data can be ensured best through a multistep review process involving one or more technical reviews, regulatory reviews (i.e., as part of a regulatory impact analysis), and an independent formal peer review. Note that a formal peer review process often has specified requirements in government. For example, the U.S. OMB Peer Review Bulletin (OMB, 2004) guidelines distinguish independent formal peer review of risk assessments used to guide federal food safety decisions from academic peer review mechanisms used for publication of risk assessments as journal articles in the scientific literature and from public comment. Independent formal peer review involves having risk assessment reports, models and underlying data, and data analyses reviewed by a specified number of independent scientists, modelers, and statisticians to ensure that the risk assessment receives a thorough, objective, and unbiased evaluation. Federal agencies are required to inform the public of plans to conduct peer reviews of federal risk assessments, by posting formal peer review plans on the Web and updating the information every 6 months. The peer review plan specifies the type of peer review process to be used (e.g., panel review, individual reviewer, or an alternative process), the number of peer reviewers, the charge to the peer reviewers, and the expertise of the peer reviewers expected to conduct the review. Once an agency receives input from peer reviewers, the risk assessment model and corresponding report are modified accordingly. For added transparency, the agency also posts the charge given to the peer reviewers, a list of the peer reviewers and their qualifications, and the agency's response to peer reviewers'

Table 2 Risk management questions addressed in selected quantitative microbial risk assessments

Pathogen/product	Risk management question	Reference
Salmonella enteritidis shell eggs	What combinations of interventions along the farm-to-table continuum are most effective in reducing the risk of salmonellosis from *Salmonella enteritidis* in shell eggs?	USDA/FSIS and FDA, 1998
Salmonella spp. in egg products	How effective is one pasteurization performance standard compared with another in mitigating the risk of illness from *Salmonella* in egg products?	USDA/FSIS, 2004a (draft), 2005 (revised)
Salmonella in RTE meat and poultry products	How effective is one lethality performance standard compared with another in mitigating the risk of illness from *Salmonella* and other microbial contaminants in RTE meat and poultry products?	USDA/FSIS, 2005a
Fluoroquinolone-resistant *Campylobacter* in chicken	What is the human health impact of fluoroquinolone-resistant *Campylobacter* infections associated with the consumption of chicken?	FDA/CVM, 2001
E. coli O157:H7 in ground beef	What combinations of interventions along the farm-to-table continuum are most effective in reducing risk of illness associated with *E. coli* O157:H7 in ground beef?	USDA/FSIS, 2001
E. coli O157:H7 in intact versus nonintact beef	Do nonintact (tenderized) beef steaks and roasts pose a greater risk of illness than intact (nontenderized) beef steaks and roasts prepared using traditional cooking practices (grilling, broiling, and frying)?	USDA/FSIS, 2002
V. parahaemolyticus in raw oysters	What is the efficacy of different intervention strategies (including microbiological standards) to reduce the risk of acquiring *V. parahaemolyticus* gastroenteritis from raw oyster consumption?	FDA, 2001 (draft); 2005 (revised)
L. monocytogenes in RTE foods	Which RTE foods should receive regulatory attention to reduce listeriosis by 50%?	DHHS/FDA and USDA/FSIS, 2001 (draft); 2003 (revised)
L. monocytogenes in deli meats	Which in-plant interventions are most effective in reducing the risk of listeriosis associated with deli meats?	USDA/FSIS, 2003a
L. monocytogenes in RTE meats and poultry products	What is the most effective allocation of federal sampling resources to verify industry food safety control for *L. monocytogenes* in RTE meat and poultry products?	USDA/FSIS, 2005b
Clostridium perfringens in RTE and partially cooked meat and poultry products	How effective is one cooling (stabilization) performance standard compared with another in mitigating the risk of illness from *C. perfringens* in RTE and partially cooked meat and poultry products?	USDA/FSIS, 2005c

comments without attribution of specific comments to specific reviewers. Given the broad array of review mechanisms (technical review, regulatory review, and formal independent peer review), reviews should occur at the beginning, middle, and end of the risk assessment process. Similarly, public input should be sought in the same manner, periodically from the initiation through development and completion of a risk assessment.

Stakeholder Involvement

The level of stakeholder involvement may depend on the complexity of the risk assessment or risk to be controlled (Renn, 2005). For example, if there is great ambiguity in the interpretation of the results of a risk assessment model, a greater level of stakeholder involvement may be necessary in developing appropriate solutions to the food safety problem compared with simple or routine issues for which there is consensus among the regulatory authorities, the regulated industry, and other stakeholders about the appropriate solution.

Stakeholders (including the consumer representatives and the food industry) may have unpublished data and information useful for a risk assessment. These data may be obtained informally or formally through a "Request for Information" or "Data Call-In" published by federal agencies in the United States in the *Federal Register*. Through this process, stakeholders are informed about the risk assessment and encouraged to provide input to the risk assessment. After a draft risk assessment is issued for public comment, stakeholder input is obtained during public meetings and through submissions to the public docket. A transparent process should be used to document and respond to stakeholder comments. For example, federal agencies should describe how stakeholder comments were used to revise a draft risk assessment, including the criteria for selecting data used in the risk assessment.

Identifying and Selecting Risk Management Options

Depending on the specifics of the problem to be solved, risk managers may identify and select a single option or multiple options to be implemented simultaneously, or they may phase in a series of options over time. Examples of possible risk management options are provided in Table 3.

The particular option chosen may be driven by the need to address an immediate concern even though the risk managers acknowledge that not all of the needed information is available. As a result, food safety decisions might initially be made based on best available information and revisited at a later date. A risk management decision might also be made using preliminary risk estimates from a risk assessment that has not been fully developed. For example, following the discovery of a cow of Canadian origin testing positive

Table 3 Examples of risk management decisions

Decision	Description
Do nothing	Usually an interim decision because sufficient information is unavailable to fully understand the issue and/or possible solutions.
Conduct research	Means of gathering the information or data to better understand the nature of the issue and/or possible solutions.
Consumer outreach/ education	Includes messages to assist consumers issued through the Internet, radio, magazines, or brochures. Messages may be targeted to a specific subpopulation, such as information on food safety for Moms-To-Be available at http://www.cfsan.fda.gov/~pregnant/pregnant.html, or to the general public, such as Ask Karen, the FSIS Virtual Representative, an automated response system available at http://www.fsis.usda.gov/Food_Safety Education/Ask_Karen/index.asp.
Action plan	Includes the steps the agency will take to reduce food-borne illness within their regulatory purview. For example, see the *Listeria* action plans available at http://www.foodsafety.gov/~dms/lmriplan.html.
Training	Federal agencies may develop and deliver training and technical assistance for industry and food safety regulatory employees.
Guidance document	Guidance documents are documents prepared for agency staff applicants/ sponsors, stakeholders, and the public that describe the agency's interpretation of policy on a regulatory issue.
Regulation	Regulations are promulgated in the *Federal Register* and codified in the Code of Federal Regulations. They may contain provisions that will be enforced as legal requirements, or that are intended only as guidelines and recommendations, or both.

for bovine spongiform encephalopathy (BSE) in 2004, the U.S. Department of Agriculture issued an interim final rule to provide safeguards against BSE including prohibiting the use of specified risk materials for human food (USDA/FSIS, 2004b). This policy, as well as others, was put into place to quickly address a public health need. The use of an interim final rule allowed time for the continued development, peer review, and then finalization of risk assessments to evaluate the risk management options and consider alternative public health measures prior to finalization of the rule.

PUTTING RISK ASSESSMENTS TO WORK

A risk assessment may be initiated by risk managers for various reasons: to collect and evaluate information, facilitate communication among affected groups, assist in the understanding of complex processes, provide a tool to evaluate proposed risk management strategies, or identify research needs to fill knowledge gaps (Fazil et al., 2005). Additional uses for microbial food safety risk assessments include the following (Buchanan, 2006): setting priorities, identifying steps that are "major contributors" to risk, evaluating the effectiveness of potential control measures, evaluating the equivalence of different control measures, evaluating the effectiveness of proposed standards and criteria, evaluating the contribution of compliance to risk, determining subpopulations "at increased risk," and assessing uncertainty and variability.

The type or design of the risk assessment needed depends on the specific risk management question or problem to be solved. Microbial risk assessments may be based on a single pathogen and food product, multiple pathogens and a single food product, or a single pathogen in multiple products. The risk assessment might follow a food from the farm, through manufacturing and production, to the consumer's table, or it could begin at any intermediate step in the farm-to-fork continuum. The case study below provides details of example "product pathway" and "risk ranking" type of risk assessments. A "product pathway" risk assessment is useful in particular for identifying factors that influence the risk and for evaluating the effectiveness of potential mitigation strategies. A "risk ranking" type of risk assessment allows the relative comparison of multiple foods or hazards and can be very useful for setting priorities. Other types of risk assessments include "risk-risk," which evaluates the overall impact of substituting one risk for another, and "geographical," which examines public health impact from the introduction of a hazard into a food system (Council for Agricultural Science and Technology [CAST], 2006).

Once a risk assessment model is developed, it can be used to evaluate the potential effectiveness of different risk management options or control measures. This evaluation has been referred to as conducting "what if" scenarios. Scenarios are conducted by changing the model input parameters and measuring the change in the model output predictions. Modeling specific scenarios can assist in the interpretation of complex risk assessments by allowing a comparison of baseline calculations with new situations. "What if" scenarios can be used in many applications and can be classified under three main types: pragmatic, heuristic, and abstract (Carrington et al., 2004) as described below.

Pragmatic Scenarios

Models that predict what may reasonably be expected to happen can also be used to evaluate the consequences of a regulatory action. A model may be used to demonstrate the changes anticipated in the food system based on the specifications of a current or proposed regulation. For example, if a regulation requires an intervention that yields a certain log-reduction level of a pathogen in a product, the model parameters can be altered by using a lower or higher log-reduction level. The information obtained can inform the risk managers of the likely impact on public health by the application of the lower or higher log-reduction intervention on the product.

Heuristic Scenarios

A scenario can also be used to evaluate the impact of further research. By altering the uncertainty distribution in the model, the impact of additional research on the model predictions can be evaluated. The need for new research or data collection can be justified if it is determined that narrowing the uncertainty distribution of that specific model parameter has a significant effect on the model predictions.

Abstract Scenarios

Scenarios can also be devised to evaluate the impact of theoretical interventions on the model outputs. These are different from pragmatic scenarios because the intervention studied may not actually exist or the changes to the baseline model may be purely academic and not be based on a known relationship. This type of scenario is useful to simulate the development of potential food safety control strategies.

Examples of pragmatic "what if" scenarios from the DHHS-FDA/USDA-FSIS 2003 *L. monocytogenes* in RTE foods risk assessment are provided below. The *L. monocytogenes* risk assessment (LMRA) identified five factors that affect potential consumer exposure to *L. monocytogenes* and the risk of illness from that exposure: (i) the amounts and frequency of consumption of RTE food; (ii) the frequency and levels of *L. monocytogenes* in food; (iii) the potential of the food to support growth of *L. monocytogenes* during refrigerated storage; (iv) refrigerated storage temperature; and (v) the duration of refrigerated storage before consumption. Growth of *L. monocytogenes* in a particular type of food is a function of the food matrix, storage time, and storage temperature. Scenarios were conducted to evaluate the potential reduction in cases of listeriosis by limiting the growth of *L. monocytogenes* that could occur in a food between sale at retail and consumption through control of home refrigerator storage temperature and time.

Refrigerator temperature scenario

The "what if" scenario considered the impact on the predicted number of cases of listeriosis by ensuring that home refrigerators are not operated above the recommended level of 5°C (41°F). The baseline risk assessment model used a distribution of home refrigerator temperatures (Audits International, 1999). For this scenario, the distribution of refrigerator temperatures was truncated, meaning that we assumed that refrigerators in the United States were operated in a manner to keep food sufficiently cool. The baseline and truncated distributions are illustrated in Fig. 3. The scenario demonstrated that the predicted number of cases of listeriosis would be reduced approximately 69% (from 2,105 to 656) by ensuring that all home refrigerators operated at

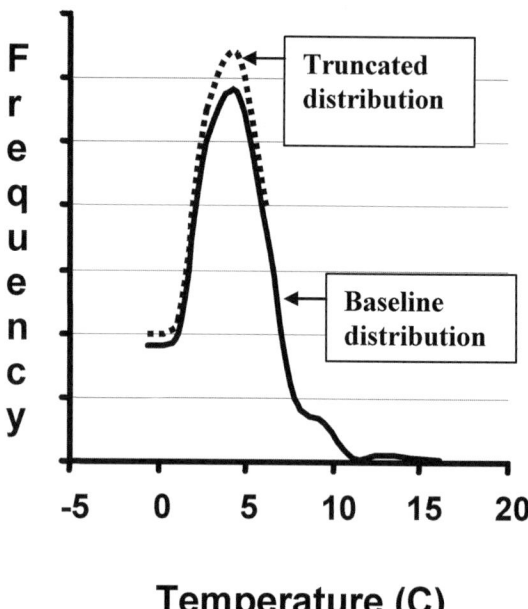

Figure 3 Baseline and truncated distributions of the frequency of home refrigerator temperatures used in "what if" scenarios for the LMRA (DHHS/FDA and USDA/FSIS, 2003).

temperatures of 45°F or less. The predicted number of cases was further reduced to 28 per year (>98%) when the distribution of home refrigerator temperatures did not exceed 41°F. This information was subsequently used in risk communication messages to consumers.

Storage times scenario
Because *L. monocytogenes* growth is also a function of the length of time a food is stored before consumption, this scenario evaluated the impact of reducing the maximum storage time (e.g., by labeling food with "consume by" dates) on the number of cases of listeriosis. Reducing the storage time for deli meat, for example, from the maximum time used in the baseline model (28 days) to 14 days reduces the predicted median number of cases of listeriosis in the elderly population from 228 to 197 (13.6% reduction). Shortening storage time to 10 days further reduces the cases to 154 (32.5% reduction).

CASE STUDY

DHHS/FDA and USDA/FSIS *L. monocytogenes* and FDA *V. parahaemolyticus* Risk Assessments

This case study is a discussion of the similarities and differences between the construction and use of the *L. monocytogenes* (DHHS/FDA and USDA/FSIS,

2001, 2003) and *V. parahaemolyticus* (FDA, 2001, 2005) risk assessments for risk management decision making. The two risk assessment efforts are compared on the basis of the type of organism studied, the food(s) of interest, characteristics of the disease, the complexity of the issue, and how these factors influenced the level of public participation, as well as the different types of risk management options considered and recommended for implementation. Also discussed are how these factors and the available data influenced the modeling techniques used and key results.

Trigger for risk assessments

Interest in conducting the *V. parahaemolyticus* Risk Assessment (VPRA) was triggered by the two outbreaks in 1997 and 1998 in the United States with more than 700 cases of *V. parahaemolyticus* illness. These outbreaks appeared to be associated with apparently low levels of *V. parahaemolyticus* in raw oysters, which raised concerns about the adequacy of the existing FDA guidelines of no more than 10,000 *V. parahaemolyticus* cells/g of oyster meats. As a result, FDA recognized the need to evaluate the factors that contribute to *V. parahaemolyticus* illnesses from consumption of raw oysters and the efficacy of different intervention strategies to reduce the risk of acquiring *V. parahaemolyticus* gastroenteritis.

While the outbreaks contributed to the overall need for the LMRA, the primary force driving the development of the LMRA was the need to achieve the goal of reducing listeriosis by 50% by 2005 as directed by the Healthy People 2010 initiative. The Centers for Disease Control and Prevention's (CDC's) FoodNet data indicated that the incidence of *L. monocytogenes* infection had decreased between 1996 and 2001 and then reached a plateau, indicating a need to more effectively focus the U.S. regulatory programs. The LMRA was designed to identify which foods should receive regulatory attention to improve public health and to address risk management options that could be applied by both FDA and FSIS.

Outbreaks associated with *L. monocytogenes* typically represent a breakdown in food production, manufacturing, or distribution systems that have been put in place to prevent *L. monocytogenes* contamination. For example, outbreaks of listeriosis have been linked to failure to protect a frankfurter-processing line from environmental contamination caused by plant renovations (1998–1999), use of defective processing equipment in the production of chocolate milk (1994), and inadequate pasteurization of milk used to make Mexican-style soft cheese (1985). However, the number of illnesses attributed to documented outbreaks is a small proportion of the total 2,500 estimated annual cases. The balance of these cases are sporadic and may reflect the occasional, chance contamination and proliferation of *L. monocytogenes* in the food in a food safety system that is otherwise "under control."

Risk assessment approaches

The approaches to the design of the LMRA and VPRA were quite different because the purpose and risk management questions to be answered were different.

VPRA is a "product pathway" risk assessment. This type of risk assessment is particularly useful for identifying factors that influence risk and evaluating the effectiveness of potential mitigation strategies. In this case, the VPRA examined the factors that influence the risk associated with *V. parahaemolyticus* in raw oysters in the pathway from harvest through postharvest handling to consumption. The VPRA was conducted under the auspices of a single agency, FDA. The scope of the VPRA was complicated by the need to assess risk resulting from each step of the food system from harvest to consumption, which also had to account for the differences in the risk of the product obtained during each season of the year from multiple harvest regions. The levels of *V. parahaemolyticus* likely to be present in raw oysters at the time of consumption were examined. Levels of *V. parahaemolyticus* in oysters are influenced by the harvest methods and environmental conditions, as well as the handling of oysters after harvest. These practices and conditions differ by geographic area and time of the year. Therefore, a model was constructed to predict illnesses for each harvest region and season in the United States and to estimate the likelihood and severity of illness following exposure to pathogenic *V. parahaemolyticus*. Risk of illness was determined for six harvest regions during the four seasons of the year. The risk assessment model was constructed to link the predicted levels of *V. parahaemolyticus* in oysters at harvest, postharvest, and at consumption to potential illnesses. Once the baseline model was developed, "what if" scenarios were run to understand the potential impact of intervention strategies to reduce both levels of *V. parahaemolyticus* in oysters and illnesses.

The product pathway approach used for the VPRA was not appropriate for the LMRA. The LMRA is a "risk ranking" risk assessment in which multiple products were evaluated. This type of risk assessment is used for priority setting to ensure that products with the greatest risk are receiving the most attention. A large variety of RTE foods were considered in the LMRA, some regulated by FDA and others by FSIS. As a result, the LMRA was a joint effort led by FDA in collaboration with USDA-FSIS and in consultation with the CDC. The involvement of multiple agencies introduced the need for a coordinated effort to reach consensus on the approach of the risk assessment. The scope of the risk assessment, however, was limited to consideration of the RTE foods from retail to consumption because the risk management questions it posed did not require a consideration of the complexities of how each RTE food is produced.

The LMRA was designed to rank the relative risk of listeriosis from consumption of different types of foods. This required estimates of what consumers are ingesting, but most of the available data were for *L. monocytogenes* levels in foods collected at retail not at the time of ingestion. Thus, the starting point of the model was the levels of *L. monocytogenes* in foods at retail. Estimates of the levels of *L. monocytogenes* in food at consumption, for each of the food categories, began with data that reflected initial distributions of *L. monocytogenes* in food at the retail level and then considered the potential for growth or decline of *L. monocytogenes* in food during transportation, home storage, and handling before consumption. These estimates were then used along with a dose-response model to quantitatively estimate the proportion of illnesses and deaths attributed to each food category. The resulting risk estimates were expressed on both an individual basis (i.e., risk per serving) and population basis (i.e., risk per annum) for each population. These predictions were interpreted and used to estimate the relative risks among the food categories. "What if" scenarios were also run to determine the influence of selected model parameters, such as home storage time and temperature, on the predicted risk. Completion of the risk ranking model was used to identify high-risk food products that would benefit from further study, such as a product pathway risk assessment (similar to the VPRA).

Organism

The organisms evaluated in the VPRA and LMRA are recognized worldwide as a significant cause of bacterial food-borne illness. *V. parahaemolyticus* was first isolated in the 1950s and has a relatively long history of causing food-borne illness. In contrast, *L. monocytogenes* did not gain prominence as a food-borne disease with major public health consequences until the 1980s. As discussed below, the timing of the emergence of these organisms as food-borne pathogens impacted the availability of human data for dose-response modeling.

The organisms differ with regard to pathogenicity. Not all strains of *V. parahaemolyticus* cause illness in humans; in fact, the majority of strains isolated from the environment or seafood are not pathogenic. Pathogenic *V. parahaemolyticus* strains were defined in the VPRA as strains that produce a thermostable hemolysin (TDH+), an enzyme that lyses (breaks down) red blood cells on Wagatsuma blood agar plates.

With regard to *L. monocytogenes*, numerous virulence components have been discovered and animal studies show a range of virulence among food isolates. Currently for regulatory purposes, all strains of *L. monocytogenes* are treated as if they were equally virulent and no distinction is made for relative pathogenicity. For the LMRA, the model includes a distribution for the range of strain virulence observed in the animal studies.

Foods of concern
The foods of concern evaluated in the VPRA and LMRA were different. Raw oysters were selected for the VPRA for several reasons. *V. parahaemolyticus* is a naturally occurring organism in coastal marine waters and estuaries, and as a result, it is found in many types of seafood, i.e., fish, crustaceans, and molluscan shellfish. It multiplies and colonizes in the gut of filter-feeding shellfish such as oysters, clams, and mussels. First, shellfish are a concern because they are handled after harvest for a significant time under nonrefrigerated temperatures before refrigeration, which allows *V. parahaemolyticus* growth. Second, although *V. parahaemolyticus* illnesses have been associated with various types of cooked and raw seafood, oysters are the most common seafood associated with *V. parahaemolyticus* infections in the United States. Although cooking destroys *V. parahaemolyticus*, oysters are commonly consumed raw compared with other types of seafood and *V. parahaemolyticus* illnesses associated with cooked seafood are likely due to inadequate heating or recontamination.

In contrast, the LMRA focused on 23 categories of RTE foods. *L. monocytogenes* occurs widely in agricultural settings (soil, plants, and water) and can establish itself in a food-processing plant. Strict in-plant sanitation measures are needed in plants producing RTE products to prevent the proliferation of the organism in the plant and to prevent contamination of finished product. The manufacture/processing of some RTE foods includes a lethality step (i.e., pasteurization of milk) which inactivates *L. monocytogenes*. These foods, however, can become recontaminated with *L. monocytogenes* at later stages in the manufacturing process, such as during packaging. Other RTE foods (i.e., fresh fruits and vegetables) have no inactivation step to control *L. monocytogenes*. *L. monocytogenes* present in RTE foods is of concern because the organism has the ability to grow in food (although at different rates depending on the physical and chemical characteristics of the specific food) even at refrigeration temperatures.

Populations of concern
Both the VPRA and LMRA evaluated the impact of the risk on the total U.S. population and various at-risk subpopulations. The CDC estimates that of approximately 7,880 *Vibrio* illnesses each year in the United States, 2,800 illnesses are estimated to be associated with *V. parahaemolyticus* and the consumption of raw oysters. The overall estimated risk of progression to septicemia occurring subsequent to *V. parahaemolyticus* illness is 0.0023, or

approximately two cases of septicemia per 1,000 illnesses. For immunocompromised individuals, however, the probability of gastroenteritis progressing to septicemia is approximately 10-fold higher, with approximately 25 cases per 1,000 illnesses. Any individual can develop illness from *V. parahaemolyticus*. In most cases, the organism causes gastroenteritis of short duration and moderate severity that is characterized by diarrhea, vomiting, and abdominal cramps. However, individuals with underlying chronic medical conditions (such as diabetes, alcoholic liver disease, hepatitis) or diseases (e.g., AIDS), as well as those receiving immunosuppressive treatments (e.g., chemotherapy for cancer), are more likely to develop severe sequellae such as septicemia, a severe, life-threatening condition.

While *L. monocytogenes* is estimated to cause a similar number of illnesses as *V. parahaemolyticus* each year, the potential severity of these illnesses is greater. CDC estimates that *L. monocytogenes* causes 2,500 cases of invasive listeriosis a year, of which approximately 20% (500) are fatal. Listeriosis is considered to be predominantly food borne, but because of the long incubation time between exposure and illness, it is difficult to associate illness with a specific food and in only in a few cases is the food that caused the disease identified. With listeriosis, although public health consequences can occur in all segments of the population, serious illness is predominantly seen in specific higher-susceptibility subpopulations such as the elderly and individuals with a preexisting illness that reduces the effectiveness of their immune system. Pregnant women are in this higher-susceptibility group with perinatal listeriosis leading to in utero exposure of the fetus, fetal death, premature birth, or neonatal illness and death.

Stakeholder interest

Differences existed between the VPRA and LMRA in both the level of interest and the number of stakeholders interested in the risk assessments. The stakeholder interest in the VPRA was centered on the oyster industry. Data for the risk assessment were obtained from published and unpublished scientific literature and reports produced by state shellfish control authorities, the CDC, the shellfish industry, the Interstate Shellfish Sanitation Conference (ISSC), and FDA shellfish specialists. Stakeholder interest was particularly high in the harvest regions where the risk assessment predicted that the levels of *V. parahaemolyticus* in oysters at the time of consumption are influenced by region-specific postharvest handling practices. New research was initiated to fill identified data gaps with most of the new work conducted by FDA.

The interest in the LMRA was industry-wide; stakeholders included major corporations with brand name recognition. After the draft LMRA was issued, industries enlisted the aid of their trade associations to fill in data gaps. For example, in one case a trade association collected approximately 30,000 samples and conducted an enumeration study to determine *L. monocytogenes* levels in selected foods at the retail level, including smoked finfish, soft cheese, bagged lettuce, deli salads, and deli meats. This effort was supported in large part by FDA through a contract between the Joint Institute for Food Safety and Applied Nutrition (JIFSAN) and the National Food Processors Association (NFPA), now the Food Processors Association (FPA). There was also industry interest in reformulating products in which high levels of *L. monocytogenes* were found (e.g., hot dogs, deli salads, and deli meats) to prevent or inhibit the growth of *L. monocytogenes*.

Risk assessment procedure

Procedures common to both the VPRA and LMRA included activities such as forming teams, publicly announcing the risk assessments, gathering data, engaging stakeholders, developing and validating the model, preparing a report, seeking and addressing reviewer comments, and finally issuing the revised model and report. Once the decision was made to conduct the risk assessment, teams were formed and data collection was initiated. Data were gathered from published sources and efforts were made to obtain unpublished data from industry, academics, trade associations, and other government and international bodies. After the baseline model was developed, the impact of various interventions could be considered.

Stakeholders were involved throughout the process of conducting both the VPRA and the LMRA. Because the LMRA encompassed food categories including meat, seafood, produce, dairy products, the number of stakeholders interested in the work was greater than that for the VPRA. Informing the stakeholders in both cases occurred primarily through formal announcements in the *Federal Register*. The first announcement focused on the decision to conduct the risk assessment, described the purpose and scope, and requested submission of specific data needed for the risk assessment. Later announcements were focused on the key assumptions, data, and modeling approaches. Public meetings were also held and advice was sought at meetings of the National Advisory Committee on Microbiological Criteria for Foods, which were also open to stakeholders and the public.

Scientific experts within and outside the government peer reviewed the risk assessment model and supporting documents. Following agency approval and clearance, the draft risk assessments were issued for public comment. The public comments, newly available data, and modeling techniques were reviewed and the model was revised accordingly. Throughout the process of conducting the risk assessment regular meetings were held with risk management and risk communication advisors. Other federal agencies were informed through briefings and updates.

Risk assessment data

Both risk assessment models required data on exposure and dose-response relationships to estimate risk, but there were differences in the types of data used.

Exposure assessment data. Data required for an exposure assessment include levels of the pathogen in specific foods; changes in the levels as a result of processing, handling, or storage; growth rate of the pathogen under specific time and temperature conditions; and consumption information.

The prevalence of *V. parahaemolyticus* in oysters at consumption is influenced by the harvest methods and environmental conditions, as well as the handling of oysters after harvest. These practices and conditions differ by geographic areas and time of the year. As a result of this complexity, there were insufficient data for all of the various combinations of harvest regions and seasons. However, because the levels of *V. parahaemolyticus* in oysters are correlated with water temperature, the levels of *V. parahaemolyticus* in oysters at harvest could be predicted, in part, by using the available data on daily water temperatures. Constructing the model in this manner allowed the predictions of illnesses from consumption of raw oysters for 24 combinations of harvest regions and seasons.

Although regional and seasonal effects were not considered in the LMRA, the scope was broad, focusing on a significant portion of the RTE foods that are consumed without further cooking (e.g., cheese, deli meat) or are reheated just prior to consumption (e.g., frankfurters). Extensive qualitative (presence/absence) data were available on the presence of *L. monocytogenes* in various foods at retail, but limited enumeration data. The lack of enumeration data was likely due to the regulatory policy that focused only on the presence or absence of the organism in food, as opposed to the number. Distributions of the range of *L. monocytogenes* in specific foods at retail were constructed using the available enumeration and qualitative data. For some foods data were also available for samples collected in countries that export those foods to the United States. Data from outside the United States were considered in the model, but the information was "weighted." Weighting the data means that in model simulations the data with lower weights are selected less frequently than data with higher weighting and thus have less impact on the predicted risk values. Data from outside the United States were given lower weight and, because of changes in the industry in the 1990s to reduce *Listeria* in food, older data were also given a lower weight.

Dose-response relationship data. As noted above, *V. parahaemolyticus* was recognized as an important human food-borne pathogen two decades before *L. monocytogenes*. Human feeding studies were conducted using *V. parahaemolyticus* until the mid-1970s. These studies were used in developing the dose-response model. In contrast, about the time that *L. monocytogenes* was recognized as a human food-borne pathogen, the ethics of these types of studies in humans began to be questioned. It is unlikely that any future feeding studies with food-borne pathogens in humans would be undertaken. For *L. monocytogenes*, because there were no studies in humans, mouse studies were used to determine the shape of the dose-response curve.

Whether the dose-response curve is developed from human or small-animal studies, limitations exist that can be accounted for in the model uncertainty. Human clinical studies do not reflect the entire population because the participants are typically healthy adult males. These clinical trials are usually limited to a few volunteers due to financial and ethical considerations, and may require extrapolation from high study-dose levels to lower exposure levels of interest. When using animal models as surrogates for humans, there is a need to consider differences in animal response and human susceptibility to the pathogen and the inherent differences between mice and humans (i.e., body mass, metabolic rate, body temperature, gastrointestinal physiology, immune systems). In both human and animal studies, the effect of the food matrix (e.g., fat content) must be considered in evaluating the effective dose versus the administered dose. To accommodate limitations in both the *V. parahaemolyticus* and *L. monocytogenes* dose-response study data, the dose-response models were adjusted to be consistent with the estimate of total annual illnesses. This adjustment ensured that the model does not over- or underpredict the mean number of illnesses. Without the adjustments the model could have overpredicted the risk by orders of magnitude.

Key risk assessment results

The VPRA concluded that the probability of illness is more likely when *V. parahaemolyticus* levels in oysters are high. For example, the probability is relatively low (<0.001%) if levels are 50 *V. parahaemolyticus* cells/g but increases to 50% if the levels are 500,000 *V. parahaemolyticus* cells/g. Significant seasonal differences occur with the highest predicted number of illnesses in summer and the lowest in the winter (see Table 4). The predicted number of illnesses and levels of *V. parahaemolyticus* in oysters were influenced by regional differences related to water-temperature levels and harvesting and handling practices. The highest numbers of predicted illnesses were associated with oysters from the Gulf of Mexico.

Table 4 Predicted mean annual number of illnesses associated with the consumption of *V. parahaemolyticus* in raw oysters[d]

Region	Mean annual illnesses[a]				
	Summer	Fall	Winter	Spring	Total
Gulf Coast (Louisiana)	1,406	132	7	505	2,050
Gulf Coast (non-Louisiana)[b]	299	51	3	193	546
Mid-Atlantic	7	4	<1	4	15
Northeast Atlantic	14	2	<1	3	19
Pacific Northwest (dredged)	4	<1	<1	<1	4
Pacific Northwest (intertidal)[c]	173	1	<1	18	192
Total	1,903	190	10	723	2,826

[a]Mean annual illnesses refers to the predicted number of illnesses (gastroenteritis alone or gastroenteritis followed by septicemia) in the United States each year.
[b]Includes oysters harvested from Florida, Mississippi, Texas, and Alabama. The time from harvest to refrigeration in these states is typically shorter than that for Louisiana.
[c]Oysters harvested using intertidal methods are typically exposed to higher temperature for longer times before refrigeration than dredged methods.
[d]Source: FDA, 2005.

In most instances water temperature is overwhelmingly the primary determinant that controls *V. parahaemolyticus* levels in oysters. Levels of pathogenic *V. parahaemolyticus* in oysters at harvest were predicted using data on (i) the relationship between total levels of *V. parahaemolyticus* in oysters and water temperature, (ii) water temperature distributions, and (iii) the ratio of pathogenic to total levels of *V. parahaemolyticus* in oysters. After oysters are harvested, levels of *V. parahaemolyticus* can increase or decline in oysters during handling and storage before consumption. Once harvested, oysters are typically stored unrefrigerated for a period ranging from a few hours to more than half a day. The potential growth of *V. parahaemolyticus* in the oysters during this period of unrefrigerated holding is a function of the air temperature at the time of harvest and the length of time oysters are unrefrigerated. When the oysters are placed under refrigeration (e.g., after arrival at a wholesaler), the rate of growth slows until oysters reach a "no-growth" temperature (i.e., below 10°C) for *V. parahaemolyticus*. The length of time during which *V. parahaemolyticus* growth occurs after the start of refrigeration and the (reduced) rate of growth during this period were estimated. At a refrigeration temperature of 45°F (7.2°C), levels of *V. parahaemolyticus* decrease slowly as cells die under this storage condition. The baseline model was used to develop "what if" scenarios to evaluate the likely impact of potential intervention strategies on the exposure to pathogenic *V. parahaemolyticus* from consumption of raw oysters. Interventions, such as reducing the time from harvesting to refrigeration, limit the growth of *V. parahaemolyticus* and subsequently were demonstrated to significantly reduce the number of predicted illnesses. Postharvest measures (e.g., freezing, mild heat treatment, irradiation, ultrahigh hydrostatic pressure) that reduced levels of *V. parahaemolyticus* in oysters also reduced predicted illnesses.

The LMRA used a statistical technique referred to as "cluster analysis" to group the model outputs (predictions of risk/serving and risk/annum) for each of 23 categories of RTE foods into risk categories. The clusters were subsequently sorted into a two-dimensional matrix (see Figure 4). Risk managers were able to use the matrix to develop different approaches to controlling listeriosis based on the relative risk and characteristics of specific foods. "What if" scenarios were developed to allow comparison of the baseline calculations

Figure 4 Two-dimensional matrix of food categories based on cluster analysis of predicted per serving and per annum relative rankings from the *Listeria monocytogenes* Risk Assessment. Source: DHHS/FDA and USDA/FSIS, 2003.

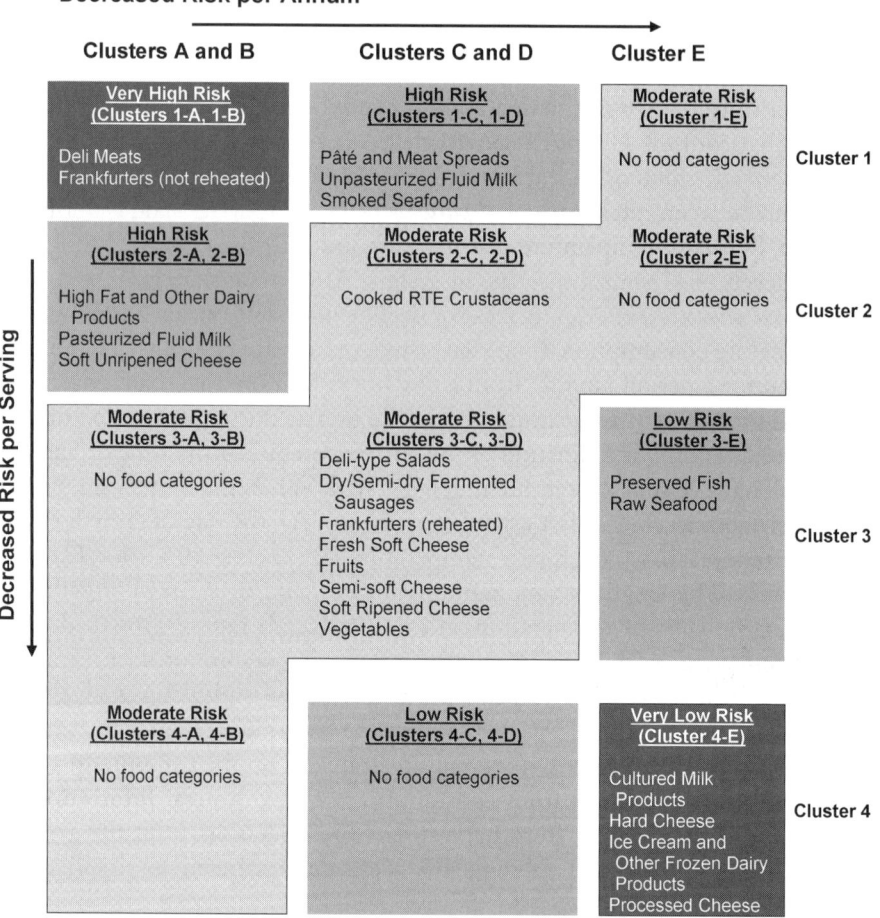

to new situations that might arise as a result of potential risk-reduction strategies. One "what if" scenario examined the impact of ensuring that home refrigerators do not operate above 41°F. In this example, the model input distribution of home refrigerator temperatures was truncated and the model rerun. With this truncation, the predicted number of cases of listeriosis was reduced from 2,105 to 28 cases per year. This scenario demonstrated the pronounced effect of proper home refrigeration on the incidence of listeriosis.

Risk management options
Microbial risk assessment is a useful tool that allows risk managers to evaluate current policy and the effectiveness of alternative options for reducing or preventing food-borne illness. For example, a risk assessment model can predict the level of a food-borne pathogen in a food under various conditions and can link that with a public health outcome.

FDA has the authority to establish limits in foods that they regulate under the Federal Food, Drug, and Cosmetic Act or Public Health Services (PHS) Act. In the case of *V. parahaemolyticus* in shellfish, FDA works cooperatively with the states, making recommendations to the ISSC, which states adopt and enforce. The ISSC is a cooperative program with industry and states established as part of the PHS Act. Currently, the FDA/ISSC recommends that the levels of *V. parahaemolyticus* in oysters not exceed 10,000 cells/g and the ISSC interim control plan (ICP) recommends monitoring of oyster meats for the presence of *V. parahaemolyticus*. Risk management options must consider that *V. parahaemolyticus* is a naturally occurring marine organism and the product in question is a raw (not cooked) oyster. *V. parahaemolyticus* is more prevalent during the summer months, and one option would be to limit the harvest at that time. The most viable options would be to limit growth, e.g., keep the level of *V. parahaemolyticus* below 10,000 cells/g through proper refrigeration and modification of harvesting methods to ensure that the product is rapidly chilled. Directed education programs for state regulators and oyster harvester programs have been adopted by the ISSC, and these have the potential to reinforce the need for adequate cooling. In addition, the benefits of potential industry changes need to be balanced with the costs or feasibility. For example, some oyster harvesters may have difficulty purchasing refrigeration capabilities for their boats but might be able to reduce the length of their harvest day. FDA and the ISSC are in the early stages of deciding how best to use the 2005 VPRA model to evaluate potential alternative means of reducing the risk of *V. parahaemolyticus* gastroenteritis associated with raw oysters.

While FDA's and FSIS's current enforcement limits for *L. monocytogenes* in RTE foods are often referred to as "zero tolerance," this is a misnomer.

The allowable level is more clearly described as "nondetectable" because the level is tied to the size of the official samples analyzed (i.e., two 25-g samples) and the sensitivity of the standard method used for *L. monocytogenes* detection in foods, which has a detection limit of approximately 0.04 CFU/g (or 1 CFU per 25-g sample). An RTE product in interstate commerce that contains detectable levels of *L. monocytogenes* is currently considered adulterated under 402(a)(1) of the Food, Drug, and Cosmetic Act in that it contains a poisonous or deleterious substance that may render the product injurious to health. If meat, processed egg, or poultry products are contaminated with *L. monocytogenes*, the products are considered adulterated under the provisions of the Federal Meat Inspection Act (21 U.S.C 601[m]), the Egg Products Inspection Act (21 U.S.C. 1033[a]), or the Poultry Inspection Act (21 U.S.C 453[g]), respectively.

The LMRA determined that RTE foods could be classified as "high-" or "low-risk" foods based on their ability to support the growth of the microorganism. The implication of this finding is that the agency could optimize the use of regulatory resources by targeting foods having the greatest likelihood of transmitting listeriosis and food processors. It also indicates that industry could reduce risk by reformulating products so that they no longer support the growth of *L. monocytogenes* or through treatment after packaging of "high-risk" foods to reduce their risk of being a vehicle of listeriosis. The LMRA identified the importance of storage temperature and shelf life of foods on the growth of *L. monocytogenes*, which enables food processors and consumers to place increased attention on these critical factors influencing health outcomes. In addition, the LMRA better characterized populations "at increased risk" of acquiring *L. monocytogenes* infection, which allows development of more refined educational messages/programs.

The LMRA was used in the development of a Listeria Action Plan (http://www.cfsan.fda.gov/~dms/lmr2plan.html), which identified areas on which to focus efforts to reduce listeriosis, including education, guidance, training, enforcement, surveillance, and research. Several options are available to further reduce the risk of food-borne listeriosis. These potential options include the following:

1. Improved control of temperature during storage and distribution
2. Improved or sufficiently stringent sanitation practices
3. Industry reformulation of products to reduce their ability to support the growth of *L. monocytogenes* or eliminate the organism in the food through cooking or pasteurization
4. Education programs for consumers to target the need to keep refrigerator temperatures at or below 40°F

The scientific evaluations and mathematical models developed for the 2003 LMRA provided a systematic assessment of the scientific knowledge needed to evaluate the effectiveness of current policies, programs, and practices and to identify new strategies for minimizing the public health impact of food-borne *L. monocytogenes*. Moreover, the assessment provided a foundation to assist future evaluations of the potential effectiveness of new strategies for controlling food-borne listeriosis. The USDA/FSIS used portions of this model in the development of a targeted risk assessment to evaluate the effects of combining testing, sanitation, and postlethality processing interventions (USDA/FSIS, 2003). The risk assessment can be used to set inspectional priorities, focus research related to intervention technologies and formulation modification, identify approaches for controlling risk, and serve as the basis for specific product/pathway analyses and risk assessments.

THE FUTURE

Risk analysis is emerging as a public health tool that promises to provide a sound scientific basis for developing national and international standards and guidelines for food safety and for supporting risk-based policies and practices to prevent food-borne illness. Risk assessment can be used to guide food safety decision making in various ways—prioritizing foods for further regulatory consideration, targeting prevention and control strategies to reduce specific food-borne illness, or allocating inspection resources toward foods that present the greatest public health risk. Although risk assessments addressing only a few food-borne pathogens have been developed to date, it is envisioned that as this tool matures, it will be used to address many more food-borne pathogens not yet considered.

Faster and Better

The future involves the continued development of risk assessments that can be tailored to provide practical guidance to risk management decisions. As additional microbiological risk assessments are developed, and the modeling techniques gain general acceptance, greater emphasis will be placed on using risk assessment models and there will be less need to explain and justify the process of initiating and conducting risk assessments. This will allow risk assessments to be developed faster and made more readily available. Increased interaction among scientists and stakeholders will allow risk assessments and models to evolve and be viewed as tools to be used today and into the future. Models developed for an industry overall can then be fine-tuned and adapted to specific production plants or manufacturers. Models will be built by connecting existing model components (modules) that can be updated more easily with new data and information as it becomes available.

To reduce model uncertainty, there needs to be greater emphasis on data collection and information management for developing risk assessments. Researchers should be encouraged to generate data specifically for use in risk assessments and to validate the models. Because a risk assessment is based on a snapshot of data available at that time, when significant changes occur in the food system, updated risk estimates should be developed and, if needed, used to refine earlier risk management decisions.

Improved Transparency

The process of developing microbial risk assessments requires ongoing public and scientific input to ensure clarity and must take into consideration stakeholder viewpoints in the development of food safety standards, guidelines, and regulations. Risk assessments and the data used to develop them can now be made more readily available to all interested parties via the Internet.

Improved Communication

As the fields of risk assessment and economic analyses become more mature, greater attention will be given to improve risk communication, in particular, in the area of communicating uncertainties in the risk estimates to decision makers, to those most at risk, and to the general public. Currently, many view risk communication as the development of consumer messages. However, development is needed of better means of communicating with and among risk managers and risk assessors. In particular, better communication of risk assessment uncertainty and variability to all interested parties, including risk managers and stakeholders, is needed. In addition, risk communicators should include an emphasis on ways to improve stakeholder involvement in the risk assessment process.

The future of microbiological food safety risk assessment is very bright. The extent of food-borne illness in the United States and continued outbreaks have highlighted the need to prevent and control food-borne pathogens at all steps in the food system from the farm to the table. Risk assessment provides a means of better linking traditional food safety techniques (e.g., microbiological sampling and testing) with public health (as measured by national food-borne disease surveillance and outbreak investigations). Microbiological risk assessments developed to date serve as powerful predictive tools to efficiently allocate resources to effectively reduce food-borne illness. However, work continues to be done to improve the use of risk assessment techniques, in general, and their use by decision makers and risk communicators.

REFERENCES

Audits International. 1999. U.S. food temperature evaluation report. [Online.] http://www.foodriskclearinghouse.umd.edu.

Batz, M. B., M. P. Doyle, J. G. Morris, Jr., J. Painter, R. Singh, R. V. Tauxe, M. R. Taylor, M. A. Danilo, and L. F. Wong. 2005. Attributing illness to food. *Emerg. Infect. Dis.* [Online.] http://www.cdc.gov/ncidod/EID/vol11no07/pdfs/04-0634.pdf.

Batz, M. B., S. A. Hoffman, A. J. Krupnick, J. G. Morris, Jr., D. M. Sherman, M. R. Taylor, and J. S. Tick. 2004. Identifying the most significant microbiological foodborne hazards to public health: a new risk ranking model. Resources for the future. Food Safety Research Consortium. [Online.] http://www.rff.org/fsrc/Discussion%20Papers/FRSC-DP-01.pdf.

Buchanan, R. L. 2006. Risk assessment for food safety and food defense. Presentation at the Food Safety World Conference and Expo, March 8, 2006. [Online.] www.foodsafetyworldexpo.com.

Buchanan, R. L., S. Dennis, and M. Miliotis. 2004. Initiating and managing risk assessments within a risk analysis framework: FDA/CFSAN's practical approach. *J. Food Prot.* 67(9): 2058–2062.

Carrington, C. D., S. B. Dennis, R. C. Whiting, and R. L. Buchanan. 2004. Putting a risk assessment model to work: *Listeria monocytogenes* "what if" scenarios. *J. AFDO* 68(1):5–19.

Center for Food Safety and Applied Nutrition (CFSAN). 2002. *Initiation and Conduct of All 'Major' Risk Assessments within a Risk Analysis Framework.* [Online.] http://www.cfsan.fda.gov/~dms/rafw-toc.html.

Clemen, R. T. 1996. *Making Hard Decisions. An Introduction to Decision Analysis,* 2nd ed. Brooks/Cole Publishing Company Inc, Pacific Grove, Calif.

Codex Alimentarius Commission (CAC). 2001. *Procedural Manual of the Codex Alimentarius Commission,* 12th ed. [Online.] www.who.int/foodsafety/publications/micro/riskanalysis_definitions/en/. Accessed 15 December 2005.

Codex Alimentarius Commission (CAC). 2003. *Working Principles for Risk Analysis for Application in the Framework of the Codex Alimentarius.* Food and Agriculture Organization of the United Nations, Rome, Italy.

Codex Alimentarius Commission (CAC). 2005. Alinorm 05/28/13, Appendix III. Proposed draft principles and guidelines for the conduct of Microbiological Risk Management (MRM), at Step 5 of the procedure. [Online.] http://www.codexalimentarius.net/web/archives.jsp?year=05.

Council for Agricultural Science and Technology (CAST). 2006. *Using Risk Analysis to Inform Microbial Food Safety Decisions.* Issue paper no. 31. CAST, Ames, Iowa.

Dennis, S., L. Carson, S. Choudhuri, and P. Klein. 2006. CFSAN's risk management framework: best practices for resolving complex food safety risks. *Food Safety Magazine* 12(1): 16–20.

Department of Health and Human Services, Food and Drug Administration and the United States Department of Agriculture, Food Safety and Inspection Service (DHHS-FDA and USDA-FSIS). 2001. Draft assessment of relative risk to public health from foodborne *Listeria monocytogenes* among selected categories of ready-to-eat foods. [Online.] www.foodsafety.gov.

Department of Health and Human Services, Food and Drug Administration and the United States Department of Agriculture, Food Safety and Inspection Service (DHHS-FDA and USDA/FSIS). 2003. Quantitative assessment of relative risk to public health from foodborne *Listeria monocytogenes* among selected categories of ready-to-eat foods. [Online.] http://www.foodsafety.gov/~dms/lmr2-toc.html.

Dyckman, L. J. 2005. The current state of play: federal and state expenditures on food safety, chap. 5. *In* S. A. Hoffmann and M. R. Taylor (ed.), *Toward Safer Food Perspectives on Risk and Priority Setting*. Resources for the Future, Washington, DC.

European Commission (EC). 2004. Science in trade disputes related to potential risks: comparative case studies. Technical report series, EUR 21301 EN, published by the European Commission, Joint Research Centre, Institute for Prospective Technological Studies (IPTS), Seville, Spain. [Online.] www.jrc.es.

Fazil, A., G. Paoli, A. M. Lammerding, V. Davidson, S. Hurdey, J. Isaac-Renton, and M. Griffiths. 2005. Microbial risk assessment as a foundation for informed decision-making: a Needs, Gaps, and Opportunities Assessment (NGOA) for microbial risk assessment in food and water. Microbial Food Safety Risk Assessment Unit, Public Health Agency of Canada. [Online.] http://www.uoguelph.ca/crifs/NGOA/Finalupdates/NGOAfinalreport.pdf.

Food and Drug Administration (FDA). 2001. Public health impact of *Vibrio parahaemolyticus* in molluscan shellfish. [Online.] www.foodriskclearinghouse.umd.edu/Vibrio.htm.

Food and Drug Administration (FDA). 2005. Quantitative risk assessment on the public health impact of pathogenic *Vibrio parahaemolyticus* in raw oysters. [Online.] www.cfsan.fda.gov/~dms/vpra-toc.html,

Food and Drug Administration, Center for Veterinary Medicine (FDA/CVM). 2001. Risk assessment on the human health impact of fluoroquinolone resistant *Campylobacter* associated with the consumption of chicken. [Online.] http://www.fda.gov/cvm/Risk_asses.htm.

Food Chemical News. 2004. Food, bug must be linked to cut foodborne illness, think-tank says. 45(52): 21–22. [Online.] http://www.rff.org/fsrc/Medis-FoodChemicalNews-02-09-04.pdf.

Hathaway, S. 1995. Harmonization requirements under HACCP-based control systems. *Food Control*. 6:267–276.

Joint FAO/WHO Expert Meetings on Microbiological Risk Assessment (JEMRA). 2006. *Enterobacter sakazii* and other microorganisms in powdered infant formula. [Online.] http://www.fao.org/ag/AGN/jemra/enterobacter_en.stm.

National Research Council (NRC). 1996. *Understanding Risk: Informing Decisions in a Democratic Society*. The National Academies Press, Washington, DC.

Renn, O. 2005. *Risk Governance: Towards an Integrative Framework*. International Risk Governance Council, Geneva, Switzerland. [Online.] http://www.irgc.org/irgc/projects/risk_characterisation/_b/contentFiles/IRGC_WP_No_1_Risk_Governance_(reprinted_version).pdf.

United States Department of Agriculture and the Food and Drug Administration (USDA and FDA). 1998. *Salmonella enteritidis* risk assessment: shell eggs and egg products. [Online.] http://www.fsis.usda.gov/Frame/FrameRedirect.asp?main=http://www.fsis.usda.gov/OPHS/risk/index.htm.

United States Department of Agriculture, Food Safety and Inspection Service (USDA/FSIS). 2001. Risk assessment of the public health impact of *Escherichia coli* O157:H7 in ground beef. [Online.] http://www.fsis.usda.gov/Science/Risk_Assessments/index.asp.

United States Department of Agriculture, Food Safety and Inspection Service (USDA/FSIS). 2002. Comparative risk assessment for intact (non-tenderized) and non-intact (tenderized) beef. [Online.] http://www.fsis.usda.gov/Science/Risk_Assessments/index.asp.

United States Department of Agriculture, Food Safety and Inspection Service (USDA/FSIS). 2003a. FSIS risk assessment for *Listeria monocytogenes* in deli meats. [Online.] www.fsis.usda.gov/OPPDE/rdad/FRPubs/97-013F/Listeriaieport.pdf.

United States Department of Agriculture, Food Safety and Inspection Service (USDA/FSIS). 2003b. Risk analysis at FSIS: standard operating procedures. *Fed. Regist.* 68:61183–61184 (October 27, 2003). [Online.] http://www.fsis.usda.gov/OPPED/rdad/FRPubs/RASOPs.pdf.

United States Department of Agriculture, Food Safety and Inspection Service (USDA/FSIS). 2004a. Draft risk assessments of *Salmonella enteritidis* shell eggs and *Salmonella spp.* in egg products. [Online.] www.fsis.usda.gov/Regulations_&_Policies/RD_04-034N/index.asp.

United States Department of Agriculture, Food Safety and Inspection Service (USDA/FSIS). 2004b. USDA issues new regulations to address BSE. News Release. [Online.] www.fsis.usda.gov/OA/news/2004/bseregs.htm.

United States Department of Agriculture, Food Safety and Inspection Service (USDA/FSIS). 2005a. Risk assessments for *Salmonella* enteritidis in shell eggs and *Salmonella spp.* in egg products. [Online.] http://www.fsis.usda.gov/Science/Risk_Assessments/index.asp.

United States Department of Agriculture, Food Safety and Inspection Service (USDA/FSIS). 2005b. Risk assessment for the impact of lethality standards on salmonellosis from ready-to-eat meat and poultry products. [Online.] http://www.fsis.usda.gov/Science/Risk_Assessments/index.asp.

United States Department of Agriculture, Food Safety and Inspection Service (USDA/FSIS). 2005c. Risk assessment for *Clostridium perfringens* in ready-to-eat and partially cooked meat and poultry products. [Online.] http://www.fsis.usda.gov/Science/Risk_Assessments/index.asp.

United States Office of Management and Budget (OMB). 2004. Final information quality bulletin for peer review. [Online.] http://www.whitehouse.gov/omb/memoranda/fy2005/m05-03.pdf.

World Health Organization (WHO). 1995a. Application of risk analysis to food standards issues. A report of the Joint FAO/WHO expert consultation. WHO, Geneva, Switzerland. [Online.] www.who.int.

World Health Organization (WHO). 1995b. Agreement on the application of sanitary and phytosanitary measures. WHO, Geneva, Switzerland.

Microbial Risk Analysis of Foods
Edited by Donald W. Schaffner
© 2008 ASM Press, Washington, D.C.

Integrating Concepts: a Case Study Using *Enterobacter sakazakii* in Infant Formula

6

Martine W. Reij and Marcel H. Zwietering

INTRODUCTION

Enterobacter sakazakii is a gram-negative, yellow-pigmented bacterium in the family of *Enterobacteriaceae*. It was previously known as "yellow-pigmented *Enterobacter cloacae*" and designated a new species *E. sakazakii* in 1980 based on differences in DNA-DNA hybridization, biochemical reactions, and pigment production (Farmer et al., 1980). Similar to many *Enterobacteriaceae*, *E. sakazakii* is considered as an opportunistic human pathogen. It has been responsible for very severe infections such as meningitis and necrotizing enterocolitis in neonates and young infants (Lai, 2001). From 1958 to 2005 a total of 45 cases of invasive infections were listed. This list comprises cases on which information was available (Food and Agriculture Organization/World Health Organization [FAO/WHO], 2006). Drudy et al. (2005) listed 72 cases of infections, for which infant formula was implicated, but not necessarily proven to be the cause.

The organism had not attracted a lot of attention in both the scientific and nonscientific press until 2001. In that year Van Acker (2001) reported an outbreak of necrotizing enterocolitis in Belgium due to *E. sakazakii* resulting from powdered infant formula. One year later the Centers for Disease Control and Prevention (CDC) reported an outbreak of *E. sakazakii* infections on a neonatal intensive care unit in Tennessee affecting nine infants. In both outbreaks case control studies suggested that a specific brand of powdered infant formula was the source. *E. sakazakii* strains cultured from patients and from the powdered infant formula were indistinguishable, confirming that contaminated powder had indeed been the source. Simultaneous with the

---

MARTINE W. REIJ AND MARCEL H. ZWIETERING, Laboratory of Food Microbiology, Wageningen University, P.O. Box 8129, 6700 EV Wageningen, The Netherlands.

publication on the Tennessee outbreak, the Food and Drug Administration (FDA) published instructions to health care professionals (see revised version at http://www.cfsan.fda.gov/~dms/inf-ltr3.html) with the aim to prevent recurrence.

Since then the number of publications on *E. sakazakii* has increased sharply. Several new microbiological methods for the enrichment and detection of the organism have been developed (for an overview see FAO/WHO, 2006) and studies on heat resistance (Edelson-Mammel and Buchanan, 2004), desiccation tolerance (Breeuwer et al., 2003), specific growth rates (Iversen et al., 2004; Kandhai et al., 2006), infectivity (Pagotto et al., 2003), and ecology (Kandhai et al., 2004) of *E. sakazakii* have been performed.

To provide scientific advice as input for the Codex Committee on Food Hygiene (CCFH) "Recommended International Code of Hygienic Practice for Food for Infants and Children," FAO and WHO convened an expert consultation in 2004. This expert consultation constructed a quantitative risk assessment model for *E. sakazakii* in powdered infant formula, suggested risk reduction strategies, and gave their recommendations to a variety of stakeholders (FAO/WHO, 2004). In 2006, a second consultation, for which a new quantitative risk assessment had been specifically designed, was asked to review the new risk assessment model, to suggest risk reduction strategies, possibly setting microbiological criteria and other control measures, and to answer a series of specific questions of the CCFH.

Risk analysis provides a framework for organizing risk management, risk communication, and risk assessment activities. Many of the activities described in this introduction so far belong to this framework. The microbiological publications provide information needed in a quantitative risk assessment; the FAO and WHO twice took the initiative to convene experts in the field to conduct a risk assessment; CCFH specified the questions to be answered (= risk management) and FAO and WHO made the outcome available to stakeholders worldwide (= risk communication). But there is more interaction between the various activities than appears at first sight. Risk communication in the form of labeling, for example, may be used as a risk management tool. And communications about the risks of *E. sakazakii* in either the scientific or the nonscientific presses have induced many researchers to work on the subject since 2002.

The recent risk assessment publications (FAO/WHO, 2004, 2006) provide a wealth of information on the ecology of *E. sakazakii*, its growth and survival in infant formula, and the influence of physical factors on the risk (see Fig. 1). Risk managers in various organizations will be able to base their decisions on this information. But one should realize that problems are seldom simple or one-dimensional. Figure 1 shows many other aspects that

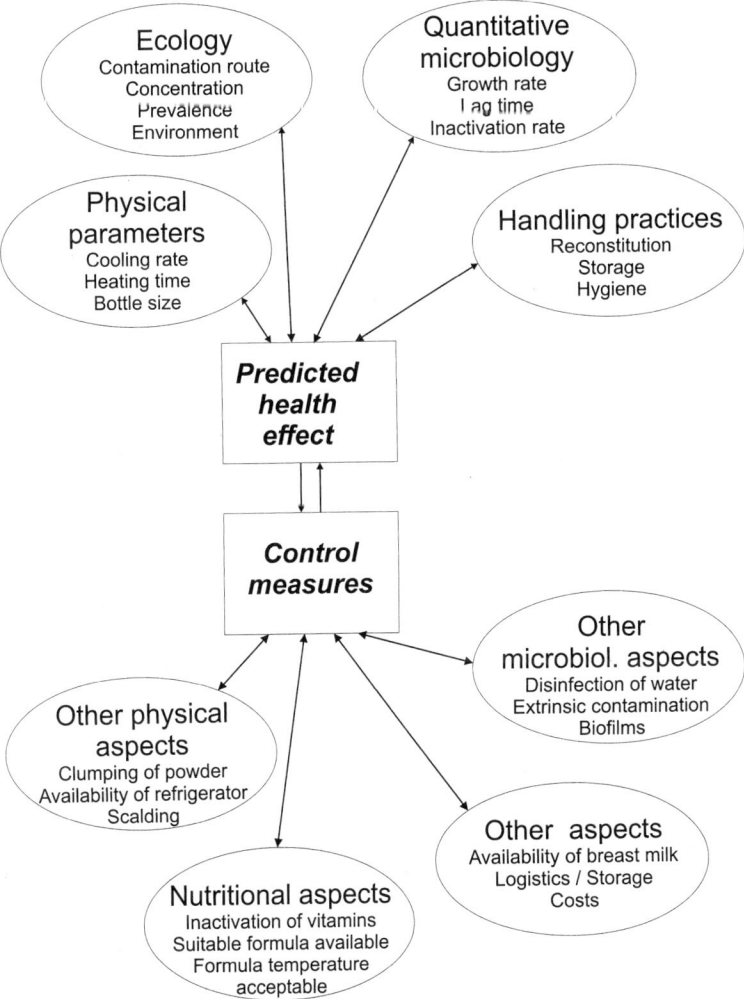

Figure 1 Examples of factors relevant for *E. sakazakii* risk assessment, risk management, and risk communication (not exhaustive).

must be integrated in the decision-making process. This chapter aims to review the risk assessment activities that have been performed on *E. sakazakii* and to focus on the relations between the various concepts.

HAZARD IDENTIFICATION

E. sakazakii has emerged as a rare cause of severe infections such as meningitis, necrotizing enterocolitis, and sepsis in neonates and young infants. A total of 45 well-defined cases of *E. sakazakii* infections were reported in the

English language literature worldwide from 1961 to 2005 (FAO/WHO, 2006). Although this list may not be complete, and although the number of *E. sakazakii* infections may be underreported, the organism can be considered to be a very rare cause of infection.

Powdered infant formula (PIF) has been implicated as the cause of infection in certain cases, but in other cases no root cause could be identified (Muytjens et al., 1983). In several cases the same strain of *E. sakazakii* was isolated from at least one patient and from opened or unopened cans of PIF (Simmons et al., 1989; Biering et al., 1989; Van Acker et al., 2001; CDC, 2002). Such isolation is strong evidence that the formula has been the source indeed. In other outbreaks *E. sakazakii* could not be isolated from the PIF, but was detected in blenders in which formula had been prepared (Noriega et al., 1990; Bar-Oz et al., 2001; Block et al., 2002).

Other food products besides PIF, such as breast milk, starch, and commercially sterilized liquid infant formula, have been described as potentially causing *E. sakazakii* infections. However, the recent expert consultation concluded that there is no evidence of a causative role of any of these foods given to young infants (FAO/WHO, 2006).

To test the magnitude of the problem of infection with *E. sakazakii* among very low birth weight (VLBW) infants (<1,500 g), blood and cerebrospinal fluid were analyzed from 10,660 VLBW infants, presumably the most vulnerable group. One case of *E. sakazakii* infection was detected. The authors conclude that outside the epidemic situation, *E. sakazakii* infection is very rare (Stoll et al., 2004).

From this conclusion one may deduct that if a case of *E. sakazakii* infection is detected in neonates or young infants, careful examination of the implicated PIF (if any) is useful. In 2004 three cases of invasive infection with *E. sakazakii* were reported within 8 days in France. Even though the patients were geographically separated, they appeared to have consumed one specific brand and type of PIF. The authorities immediately contacted the manufacturer of the formula, who on the same day withdrew the product from the market. A nationwide investigation detected one additional case of infection and five cases of colonization with *E. sakazakii*. The implicated PIF was proven to be the cause indeed. After the withdrawal no new illnesses occurred (Coignard and Vaillant, 2006). This outbreak shows that immediate association of cases of invasive *E. sakazakii* infection in neonates with the potential causative agent, a specific powdered infant formula, was warranted. The immediate withdrawal, without having any proof of a causative role at that moment, may well have prevented more infections from occurring.

Severe infections like meningitis and bacteremia are an important cause of neonatal death and account for 36% of all neonatal mortality (Lawn et al.,

2005). This chapter on the risk assessment of *E. sakazakii* is a clear example of focusing the research and attention on a relatively small risk recognized in rich countries, while 99% of neonatal deaths occur in low-income and middle-income countries (Lawn et al., 2005). The section on management options in this chapter will pay attention to the options that can reduce the risk worldwide.

HAZARD CHARACTERIZATION

E. sakazakii has been shown to cause severe invasive infections in neonates (<28 days) and infants (<1 year). The organism may affect the intestines (causing necrotizing enterocolitis), or invade the central nervous system (causing cerebritis and/or meningitis), or invade the blood (bacteremia and/or sepsis). In particular, premature babies and those born with a low (<2,500 g) or very low (<1,500 g) birth weight are considered to be at risk.

Recent analysis of 45 cases (Bowen and Braden, 2006) suggests that distinct groups of patients suffer from invasive illnesses: term infants (babies born after at least 37 weeks of gestation) developing meningitis often before the age of one week and mostly premature infants developing bacteremia after a median period of one month. Outcome differs between types of infection. Meningitis has a high mortality rate and many of the survivors (74%) suffer from severe neurological disorders and retarded development. Infants suffering from bacteremia tend to fare better; mortality among these cases is 10%. Both types of infections have occurred both in hospitals and at home, and the latest case of bacteremia was reported at as late as 10 months of age in an immunocompromised infant. This overview shows that not only premature and (very) low birth weight infants are at risk. In fact, all infants (<1 year) should be considered at risk, with neonates and those less than 2 months of age at greatest risk (FAO/WHO, 2006).

Note that only patients suffering from bacteremia and meningitis have been included in the overview. Other end points of concern, such as necrotizing enterocolitis (NEC), were not listed (Bowen and Braden, 2006) but do exist. In an outbreak in Belgium, for example, 12 infants within or shortly after their neonatal period suffered from NEC due to *E. sakazakii* infection. One of the infants died of NEC, another died of bacteremia (and was included in the overview of 45 cases), while the others recovered (Van Acker et al., 2001). The group of infants at risk for NEC, however, seems to be similar to the group defined to be at risk for bacteremia and meningitis.

A few infections have been reported in children older than 12 months (3 cases) and adults (8 cases). These infections were nosocomial and the adult patients had severe underlying diseases (Lai, 2001). To our knowledge, there is no evidence of food-borne *E. sakazakii* infections in patients over 1 year of age.

To characterize the group at risk further, note that various groups of infants receive their nutrition in different ways. Most infants are fed their mother's milk. In a hospital setting premature infants and VLBW and LBW infants might receive supplements with the breast milk because of their specific nutritional requirements. When the mother chooses not to or cannot breastfeed, infants are fed infant formula from a bottle or cup. The children born to human immunodeficiency virus (HIV)-positive mothers are very important in this respect. As the HIV virus can pass from the mother to her child via the breast milk, HIV-infected mothers are advised to avoid breastfeeding completely, provided that a replacement feeding is acceptable, feasible, affordable, sustainable, and safe (Newell, 2004).

Tube feeding requires special care. Many premature infants have to be fed with a gastrointestinal tube when they are not yet able to swallow. Nonpremature infants may have to be tube fed as well if they are too weak because of illness. Tube feeding may be applied continuously, with the formula hanging at room temperature for several hours. Tube feeding may also be applied periodically, feeding the infant numerous times per day with small amounts of milk or formula by using a syringe that is attached temporarily to the tube. Biofilm formation (Iversen et al., 2004) or stagnant product in such devices should be prevented. As far as we know, no quantitative data are available on the numbers of infants who receive the various types of nourishments or on the details of their applications. The variety of applications will be elaborated on in the exposure assessment section and the overview of management options.

There is little information on the dose of *E. sakazakii* that is required to become ill. It can be assumed that illness can result from 1 CFU of *E. sakazakii* per serving, though this probability is very low (single-hit scenario). The probability of illness after consuming a certain dose D can be described by the exponential dose-response model:

$$P_I = 1 - e^{(-rD)}$$

where P_I is the probability of illness, r is the exponential dose-response parameter, and D is the dose at consumption. The exponential model is a nonthreshold model that is linear at low doses ($P_I \sim r.D$). Taking into account the low level of *E. sakazakii* in PIF, one can assume that the dose-response relation is probably in the linear range. In fact, this assumption is quite reasonable as long as $D \cdot r < 0.1$ (Zwietering, 2005). Based on data on the Dutch population, the parameter r was estimated to be 8.9×10^{-6} or smaller (FAO/WHO, 2004). So as long as the dose is below 1×10^4 cells, one can safely assume that one still is in the linear part of the dose-response relation.

Following the same approach, we here estimate the relative susceptibilities among infants of different weight groups. Of the 10 Dutch case reports of *E. sakazakii* infection (see Table 1), there were 2, 4, 2, and 2 patients in the categories <1,500 g, 1,500 to 1999 g, 2,000 to 2,499 g, ≥2,500 g, respectively. From these data the relative probability of illness of the various groups was calculated to be 158, 152, 29, and 1, respectively, compared with the susceptibility of the neonates with a birth weight of 2,500 grams or more (see Table 1). When analyzing the 45 worldwide cases of infection the relative susceptibilities were found to be 250, 57, 12, and 1, respectively (see Table 2), based on the assumption that the distribution of the infants over the various weight categories is similar to the Dutch distribution. Although based on a very limited number of cases these relative susceptibilities are of the same order of magnitude and corroborate the hypothesis that LBW and especially VLBW infants (<1,500 g) have an increased risk of infection. These relative susceptibilities can be of practical relevance while selecting risk reduction strategies based on quantitative risk assessment.

To determine the average dose (D) ingested by an infant, one needs to consider the total consumption for the period considered and the average concentration over all feedings; thus:

$$D = C \cdot M \cdot S$$

Table 1 Reported cases of *E. sakazakii* infections in The Netherlands over a period of 40 years and the resulting relative susceptibilities and upper-limit r-values for the various weight categories

Birth weight (g)	Cases	% of newborns	Relative susceptibility	Estimated r-value
<1,500	2	0.6	158	1.84×10^{-3}
1,500–1,999	4	1.25	152	1.38×10^{-3}
2,000–2,499	2	3.25	29	2.38×10^{-4}
≥2,500	2	94.9	1	4.53×10^{-6}
Total	10	100		

Table 2 Cases of *E. sakazakii* infections reported worldwide over a period of 40 years and the resulting relative susceptibilities for the various weight categories

Birth weight (g)	Cases	% of newborns	Relative susceptibility
<1,000	10	0.6	250
<1,500	9		
1,500–1,999	9	1.25	57
2,000–2,499	5	3.25	12
≥2,500	12	94.9	1
Total	45	100	

where C is concentration of *E. sakazakii* in the powder (CFU/g), M is mass per serving (g), and S is number of servings in the period considered.

We assume that 25% of the Dutch neonates receive powdered infant formula (http://www.lalecheleague.org/cbi/bfstats03.html. Accessed 3 August 2006) and consider each infant to be at risk during the total neonatal period (28 days). The average concentration is assumed to be -3.84 log CFU/g (FAO/WHO, 2006) and the consumption volume is 150 ml of liquid formula per kg body weight per day. For VLBW infants (1,000 to 1,500 g) the daily consumption is 1.4 kg of BW × 150 ml of liquid/(kg of BW) = 210 ml. This corresponds to 21 g of powder = 7 servings of 3 g of powder/serving.

The cumulative dose (D) for one VLBW infant over the neonatal period of 28 days is calculated as:

$$D = C \cdot M \cdot S = 10^{-3.84} \text{ (CFU/g PIF)} \times 7 \text{ (servings/day)}$$
$$\times 3 \text{ (g PIF/serving)} \times 28 \text{ days}$$
$$= 0.085 \text{ (CFU)}$$

So not all babies are exposed to the organism following this calculation, even in the total period of 28 days. Note that this dose is calculated assuming that there is never any growth, so the dose is a minimum value.

In the group of VLBW infants there have been two cases in 44 years. Per year 193,750 babies are born, of which 0.6% are in this weight group (=1,162 babies) of which 25% are assumed to be fed with PIF (= 291 babies). So the risk per baby (PI) in this weight group is P_I = 2 (cases/44 years)/291 (babies/year) = 1.56×10^{-4} (cases/baby). The P_I value for this weight group is very similar to the 1 of 10,660 VLBW infants found to be infected with *E. sakazakii* (Stoll et al., 2004).

Now the r-value can be estimated to be smaller than:

$$r < P_I/D = 1.56 \times 10^{-4} / 0.085 = 1.84 \times 10^{-3}$$

Table 1 shows that r-values over the neonatal period vary from $<1.84 \times 10^{-3}$ for the VLBW category to $<4.53 \times 10^{-6}$ for the largest group of infants over 2,500 grams based on the Dutch data. The values are upper values of r since in reality some growth will have occurred, meaning a higher dose (D) and thus a smaller r-value.

EXPOSURE ASSESSMENT

To assess the probability of infection with *E. sakazakii*, the whole chain from cow to child needs to be followed, i.e., from the raw milk and the other ingredients, via the factory, the reconstitution of the powder, to the feeding of the child. An exposure assessment requires that the process is described as accurately and in detail as possible. Figure 2 shows a schematic view of the production of the formula and Figure 3 outlines several of the methods of handling and storage that have been observed. Unlike a HACCP study, in which one specific process or process line is described, a risk assessment should cover all processes with their variations.

Both Fig. 2 and 3 will be discussed in detail below and the numbers in the text indicate the processing steps as shown in Fig. 2. We have tried to include all the relevant options, but the overview might not be complete. The risk

Figure 2 Schematic view of infant formula production with various options to add ingredients.

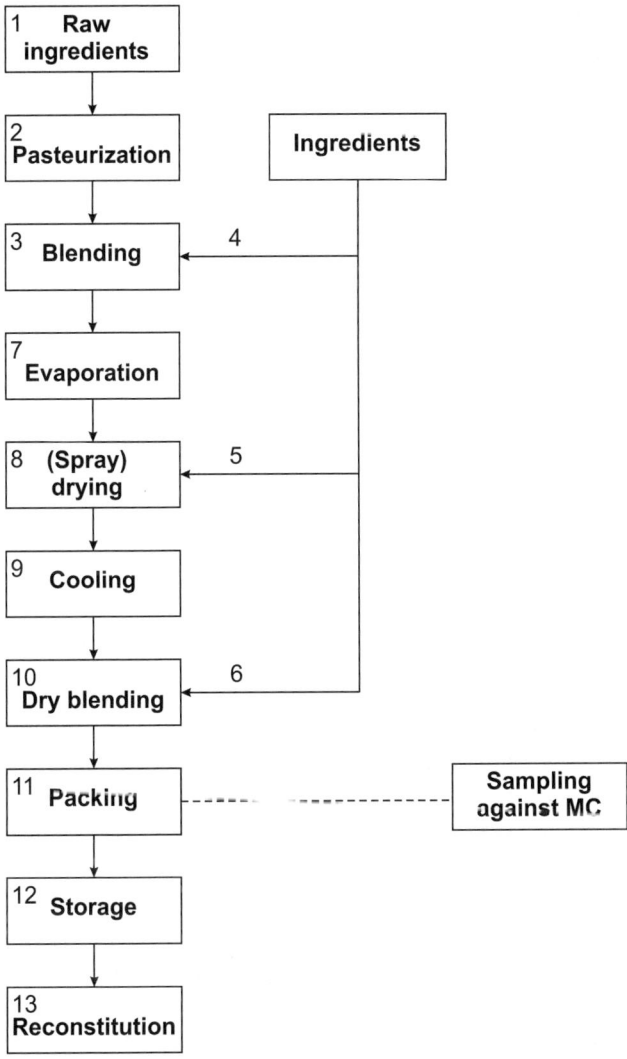

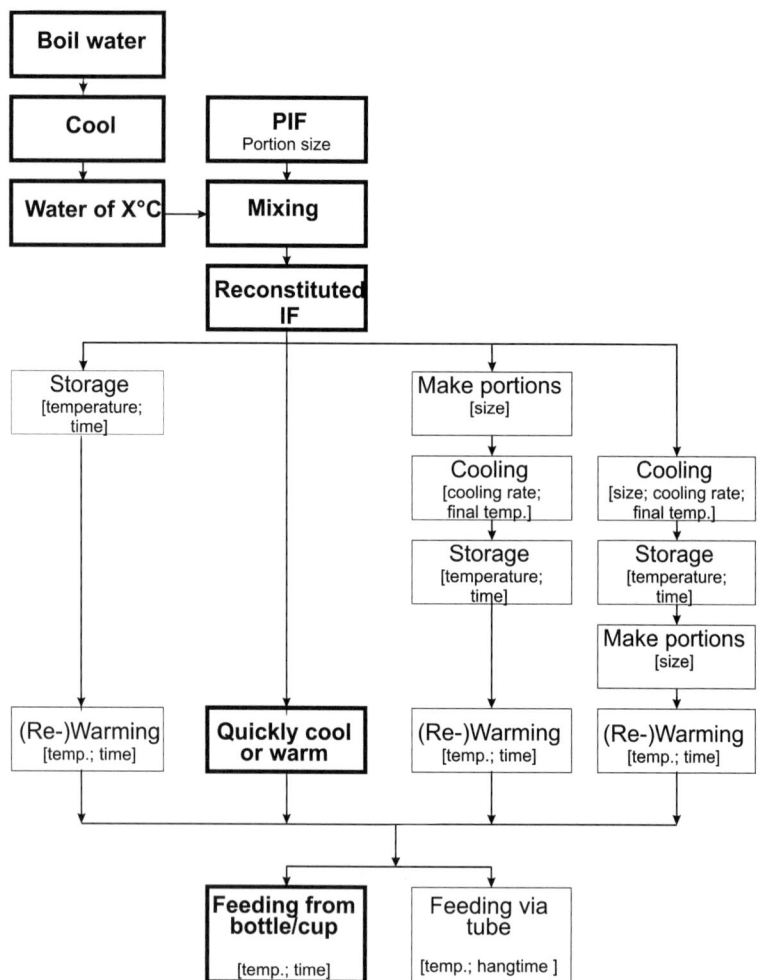

Figure 3 Overview of various methods of handling, storing, and using PIF.

assessment tool recently developed for the second expert consultation on *E. sakazakii* allows simulating many variations estimating the effect of changes and control measures on relative risk (FAO/WHO, 2006). Here we discuss various features and assumptions of this risk assessment model, further referred to as RAModel.

Contamination with *E. sakazakii* may in principle occur via its presence in raw milk or other raw ingredients (step 1). Pasteurization of these ingredients (step 2), however, results in a reduction of *Enterobacteriaceae* including *E. sakazakii* of at least 10 log-units. Its presence in the processed product is more likely to be due to contamination after the heat treatment. Dry infant formula is not a sterile product. Small numbers of bacteria, including *E. sakazakii* and other *Enterobacteriaceae*, may occasionally gain access to the processing environment and to processing equipment and may contaminate the powder via that route, notably during cooling (step 9) or packing (step 10). Other potential sources of intrinsic contamination are the heat-sensitive ingredients such as essential vitamins that are added during blending (step 4), spray drying (step 5), or mixed into the dry base powder (step 6) without further heating (FAO/WHO, 2006).

Contamination may also occur during handling and preparation of the powder in the home or in hospital settings. This type of contamination is called extrinsic contamination. It has been implicated in several outbreaks, where blenders or other equipment were proven to be heavily contaminated (Noriega et al., 1990; Bar-Oz et al., 2001; Block et al., 2002). These blenders might have been contaminated by earlier use with other contaminated powders or by other environmental sources of the bacteria. Environmental sources are quite probable. Although *E. sakazakii* is not able to form spores, it was isolated from numerous dry food production environments and from households (Kandhai et al., 2004). In the RAModel as discussed during the latest expert consultation in January 2006, only intrinsic contamination is considered. Extrinsic contamination was not taken into account (FAO/WHO, 2006).

One of the potential control measures to reduce the risk of *E. sakazakii* is setting microbiological criteria (MC) for PIF. When a microbial criterion is in place, batches of product need to be sampled at the end of manufacturing, e.g., at packaging, prior to storage (step 11 in Fig. 2). To estimate the effect of a criterion on the risk reduction, the concentrations of *E. sakazakii* in the product need to be known. These concentrations were estimated from prevalence data provided by industries, assuming that sampling is a Poisson process. By doing so, *E. sakazakii* concentrations were categorized to be in one of the following ranges:

- $-5 \log \text{CFU/g}$ (0.00001 CFU/g = 1 CFU/100 kg)
- $-4 \log \text{CFU/g}$ (0.0001 CFU/g = 1 CFU/10 kg)
- $-3 \log \text{CFU/g}$ (0.001 CFU/g = 1 CFU/1 kg)

The RAModel includes a module to evaluate risk reduction associated with setting MC for *E. sakazakii* and for *Enterobacteriaceae*. The model estimates the different relative risk reductions achieved and the percentage of product lots rejected (FAO/WHO, 2006). A selection of results is shown in the Effects of Microbiological Criteria section.

Powdered infant formulae have a very low water activity (<0.25), allowing no microbial growth of any kind during storage (step 12). *E. sakazakii* has been shown to survive for prolonged periods in the dry powder (Edelson-Mammel et al., 2005). The concentrations in PIF decline very slowly over time. The risk assessment model assumes that numbers decline at the rate of 0.001 log units per day. Because this rate is very slow, the effect of dry storage on the numbers appears to be very limited (FAO/WHO, 2006).

Figure 3 presents an overview of several different ways in which powdered infant formulae can be handled and used for infants in homes or health care institutions. As already discussed in the Hazard Characterization section, infants are fed in numerous ways, either from a bottle or cup or by continuous or intermittent tube feeding. As reconstituted infant formula is a very good medium for growth of *E. sakazakii* and other *Enterobacteriaceae*, a number of preparation and storage practices may increase the risk for infection.

Which procedures are actually followed worldwide during preparation, handling, and storage of infant formula is not clear. Producers generally advise to boil water, cool it to 40 to 50°C, add the PIF, mix, and use immediately after checking the temperature. Remaining formula should be discarded and prepared formula should not be stored. This procedure is outlined in bold in Fig. 3. In practice, however, all kinds of variations are applied, but the evidence is merely anecdotal. As an example, PIF was reported to be prepared in batches of 25 liters that were stored until use. But variations in handling do not necessarily increase the probability of bacterial growth. Hospitals, for example, were observed to prepare formulae with sterile water at room temperature or with refrigerated sterile water, which may rather decrease the probability of bacterial growth (unpublished results).

The lack of information may be due to gaps in communication between the risk assessors and the large number and wide variety of stakeholders in the field. Preceding the building of the RAModel, experts knowledgeable in the field of infant nutrition worldwide were asked for their experiences with preparation and handling of PIF. Their answers indicated a wide variety of practices. But their answers were anecdotal and mostly lacked necessary details on temperatures of reconstitution, batch size, holding times, and feeding times. To our knowledge, such data have never been gathered systematically.

The solution that was found for the absence of information on handling practices was to test and compare many options and to ultimately publish the RAModel, so that everyone could fill in his/her own parameters on preparation time/temperature, storage time, temperature, and so on. The RAModel allows testing various scenarios by dividing the process into four stages, the parameters of which can be adapted separately:

- Reconstitution of the powder
- Cooling or holding of prepared formula prior to feeding

- Warming of formula in preparation for feeding
- Feeding of the infant

In each stage the user can adapt the duration, the ambient temperature, and the rate at which the formula is heated or cooled (FAO/WHO, 2006). A selection of results is shown in the risk characterization section of this chapter. The tool itself will eventually be made available by FAO and WHO along with the publication of the expert consultation report.

Data on the growth of *E. sakazakii* at various temperatures (see Fig. 4), its heat resistance, and the cooling rates of various types of containers and bottles (Nazarowec-White and Farber, 1997; Iversen et al., 2004; Kandhai et al., 2006, Kandhai et al., in preparation) are available. With these data and the assumptions on handling practices, the temperature profile of the infant formula after reconstitution was estimated. An example is given in Fig. 5. As inactivation rates, growth rates, and lag times at each temperature can be estimated based on published experimental data, the resulting increase or decrease in bacterial count can be assessed.

Figure 4 Square root of measured and fitted specific growth rates as function of the temperature. Bullets represent growth rates published by: ◆, Kandhai et al. (2006); +, Nazarowec-White and Farber (1997); ◊, Iversen et al. (2004). The line represents a fit by the secondary-growth model of Rosso (fitted to square root of transformed data of the first publication only). The resulting parameter values are: T_{min} = 3.60°C; T_{max} = 47.6°C; T_{opt} = 39.4°C; μ_{opt} = 2.31 h^{-1}. Adapted from Kandhai et al. (2006).

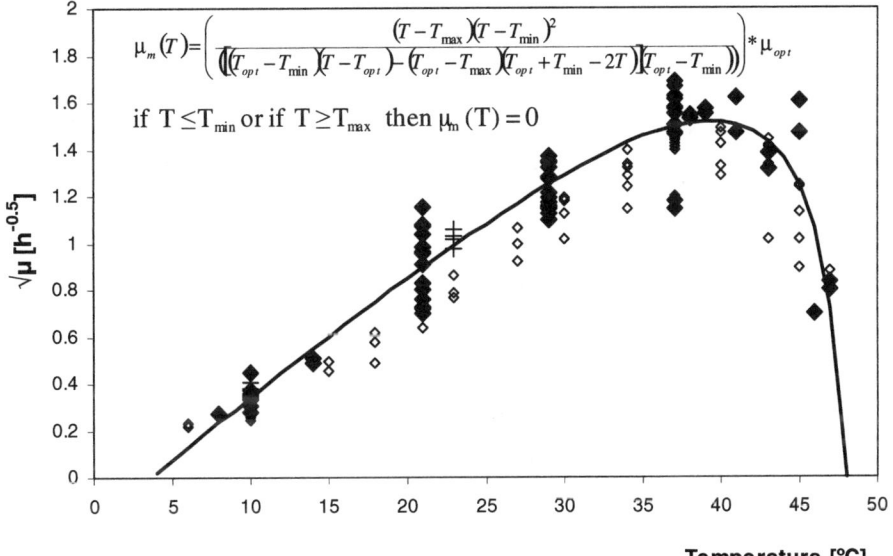

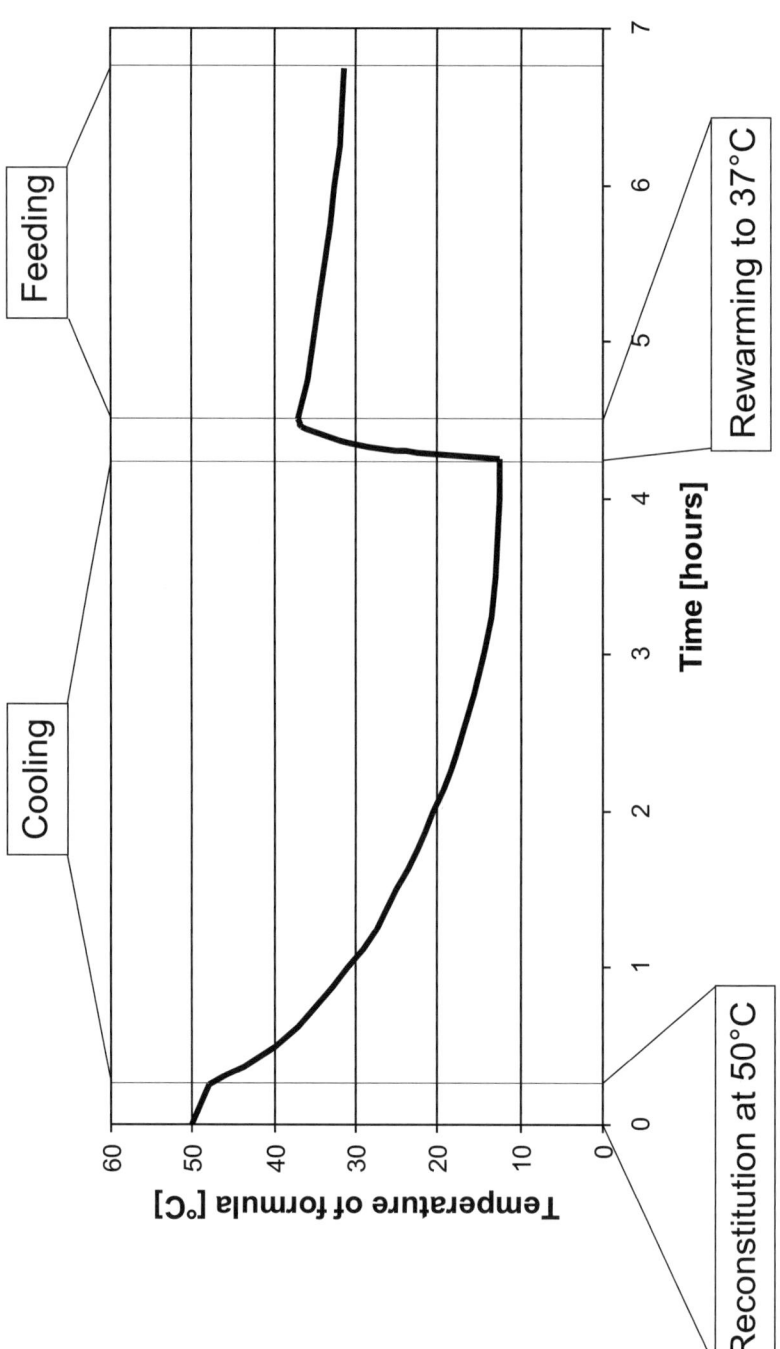

Figure 5 Example of a temperature profile that was generated by the RAModel (FAO/WHO, 2006) based on a scenario for reconstituting PIF for 15 min, cooling the bottle for 4 h, rewarming for 15 min, and feeding at 30°C for prolonged time (2 h).

RISK CHARACTERIZATION

Risk can be characterized in various units expressing the adverse public health events. Risk assessment end points may be expressed as individual risk (such as the risk per serving) or as population risk (such as the annual estimated number of illnesses or deaths), or both (Council for Agricultural Science and Technology [CAST], 2006). For *E. sakazakii* the number of illnesses cannot be estimated with confidence, as information on its dose-response relationship is not sufficient to accurately relate numbers of *E. sakazakii* with numbers of illness. Therefore in the recent risk assessment (FAO/WHO, 2006) two other types of end points other than public health events were defined in such a way that food safety management decisions can be based on the outcome of the risk assessment.

The first type of endpoint is the relative risk reduction that can be achieved by setting MC varying in stringency (see Fig. 6). It is assumed that each lot is sampled. The risk reduction achieved by a sampling plan is compared with the baseline risk (= no sampling). The baseline is set at 100%, representing the current cases of illness in a population per year. A sampling plan with a relative risk reduction of 2 is estimated to reduce the cases of

Figure 6 Illustration of the microbial quality distribution of various (hypothetical) categories of PIF on the market. The brands are assumed to fall into three levels of microbial quality, representing best (category A, — · — ·), moderate (category B, ———), and poor (category C, — — —). The vertical line represents an arbitrary MC. Adapted from FAO/WHO (2006).

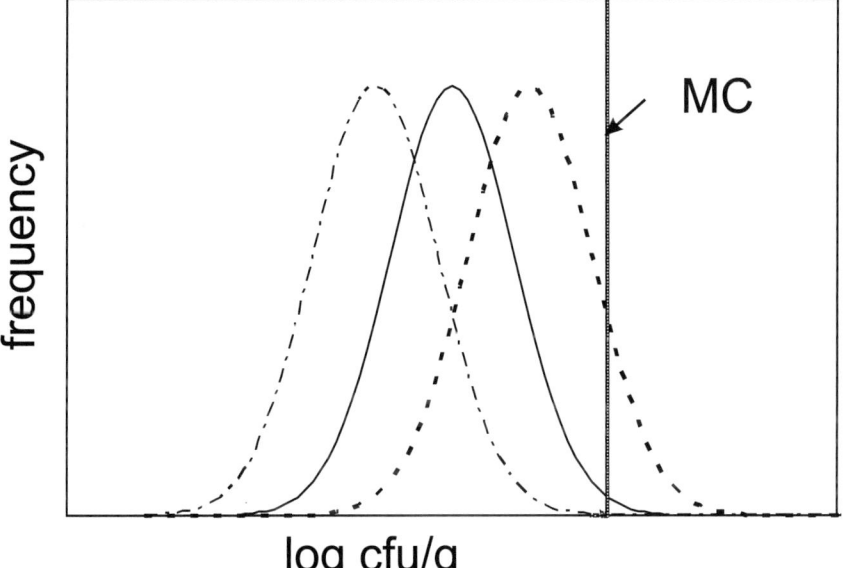

illness by a factor of 2. Apart from this relative risk reduction, the risk assessment also estimates the rejection rate, i.e., the percentage of lots that are found to have elevated levels of contamination and have to be rejected. Details on the use of risk estimates for selecting MC as a control measure to reduce the risk of *E. sakazakii* are given in the next section.

The second type of end point of the risk assessment model (RAModel) is a relative risk as well. The model predicts the level of contamination, and hence the ingested dose, resulting from feeding reconstituted PIF. The RAModel assumes that, if the product is contaminated, exactly 1 CFU is present in a contaminated serving at reconstitution, prior to any holding, cooling, warming, etc., that the serving may undergo. The level of 1 CFU per serving is logical as the concentration of *E. sakazakii* is estimated to be very low in the range of 1 CFU per kg of powder or less (see Exposure Assessment section). However, if the contamination would occur in clumps and nests of *E. sakazakii* would be present in the PIF, the assumption of 1 CFU per serving might underestimate the concentration. No information on the distribution of *E. sakazakii* is available. Depending on the handling and storage scenario (such as the scenarios shown in Fig. 3 and 5), that 1 CFU may proliferate or be inactivated. The number of cases from PIF consumption per 1 million infant-days, defined as N_{Es}, is then estimated as follows:

$$N_{Es} = \Theta \cdot C_m \cdot P_I$$

Here, N_{Es} is the number of illnesses due to *E. sakazakii* resulting from PIF consumption per million infant-days, Θ is the concentration of *E. sakazakii* in the dry product at the point of preparation of PIF, C_m is the daily PIF consumption rate per one million infants, and P_I is the probability of illness resulting from the dry powder (with 1 CFU/serving) at the time that it is consumed by the infant. During preparation and holding this 1 CFU may have increased by growth or decreased by inactivation. P_I gives the probability of illness resulting from the concentration of *E. sakazakii* at the time of consumption.

The RAModel predicts the number of illnesses due to *E. sakazakii* per million infant-days, resulting from feeding reconstituted PIF, and compares this number with the numbers resulting from a baseline scenario. Each time that a series of scenarios is calculated, one of the scenarios is chosen as the baseline scenario of that series. For this reason results should not be compared between tables on a purely numerical basis as different baselines may apply.

Using this approach numerous simulations were performed, estimating the impact of a variety of control measures. These will be described in detail in the next section.

ADVICE TO RISK MANAGEMENT: COMPARING THE OPTIONS
Preventing Intrinsic Contamination during Manufacturing

For many stakeholders the "easiest" control measure to prevent illnesses due to *E. sakazakii* in PIF is perceived to be the prevention of its occurrence in PIF altogether. Due to lack of insight in the contamination routes and lack of quantitative data on the effect of interventions, however, it was not possible to develop a risk assessment model describing the impact of interventions and control measures in the manufacturing plant. Based on data and information provided by industry, the recent expert consultation recommended the following interventions to control the contamination of PIF by *E. sakazakii* (FAO/WHO, 2006):

- Effective implementation of preventive measures as originally designed to control *Salmonella*;
- Strengthening these measures by further minimizing entry of the microorganisms and by avoiding their multiplication. To avoid duplication exclusion of water from the processing environment is essential and cleaning should be performed dry;
- The selection of suppliers of dry mix ingredients who are able to fulfill the microbiological requirements;
- Implementation of a monitoring plan targeting *Enterobacteriaceae*, as indicators for process hygiene, and *E. sakazakii* in relevant samples to demonstrate control or to detect deviations and assess the effect of corrective actions.

In this respect note that *E. sakazakii* is widely distributed and is not only present in infant formula-manufacturing plants. Kandhai et al. (2004), for instance, were able to isolate the organism from 5 of 16 households. Infant formula may be contaminated with *E. sakazakii* (and with other potential harmful microorganisms) by extrinsic sources after opening of the package of PIF, during reconstitution, or during further handling. In a previous risk assessment (FAO/WHO, 2004) extrinsic contamination was estimated to account for 20% of the cases of illness. This number was based on an overview of historic cases of illness for which the source of the contamination could be established to be PIF and on those cases where intrinsic contamination of PIF was unclear or could be ruled out. In the exposure assessment as modeled in the RAModel, extrinsic contamination is not considered.

Extrinsic contamination may also reduce the risk reduction rates as predicted from MC. Extrinsic contamination occurs during preparation, so any such event of extrinsic contamination poses an additional risk that cannot be diminished by applying MC to all batches at the end of production, before they reach the market. Consequently, the effect of the control measure is overestimated.

On the other hand, growth scenarios during preparation, storage, and handling do apply to any microorganism that is present in the formula after reconstitution, irrespective of its intrinsic or extrinsic cause. In this respect the effect of interventions as estimated by the RAModel does apply to extrinsic contamination as well. The probability of extrinsic contamination, however, and the effect of hygiene measures on that probability are not considered in the RAModel.

Effects of Microbiological Criteria

Setting a microbiological criterion is one of several food safety management tools that can be applied by governmental organizations or in international trade to reduce the risk of *E. sakazakii* infections due to contaminated PIF. If applied to each lot of PIF, MC can be used to directly identify unacceptable lots that do not comply with the established limit. Such lots cannot be released into the market to be consumed by infants.

Each microbiological criterion includes the microbiological limit that is to be implemented, the testing method to be employed, the sampling plan (i.e., size and number of samples to be examined), and the actions to be taken when the microbiological limit is exceeded (International Commission for Microbiological Specifications for Foods [ICMSF], 2002). The risk assessment model used by FAO/WHO's second expert consultation was specifically developed to test the effectivity of various sampling plans in reducing the risk of infection, by taking those lots out of the market that do not comply with an MC. Apart from the risk reduction achieved by each sampling plan, the model also estimated the rejection rate, which is the percentage of defective lots (FAO/WHO, 2006).

The stringency of an MC is determined by the microbiological limit (absence or a specified number of CFU/g), the number of samples (n), and the size of the sample (s). A criterion becomes more stringent when the limit is reduced and when the number (n) and/or the size (s) of samples increases. For *E. sakazakii* two-class attribute sampling plans were evaluated for their ability to reduce risk and their effect on lot rejection. In the sampling plans considered, only the absence or presence of the organism is relevant. If the organism is detected in one or more of the samples, the lot is rejected.

The PIF currently on the market was assumed to have a mean log concentration of either -5, -4, or -3 log CFU/g. These three categories are represented in Fig. 6. For the between-lot standard deviation (σ_b) across all PIF lots the values of 0.5 and 0.8 were simulated and for the within-lot standard deviation (σ_w) the values of 0.1, 0.5, or 0.8 were tested. Estimated risk reduction rates were simulated for sampling plans with different stringencies. A small selection of the results is shown in Fig. 7.

By application of a sampling plan according to MC, risk is reduced by taking those lots off the market that are at the right-hand tail of the distribution, surpassing the established MC (Fig. 6). The risk reduction that can be achieved, however, is limited. Figure 7a shows that the risk reduction achieved by sampling depends on the number of samples taken (n) but also on the concentration of *E. sakazakii* in PIF currently on the market. If the lot is in the best category A (-5 log/CFU or 1 CFU per 100 kg, squares in Fig. 7), none of the sampling plans tested have any notable impact on the risk

Figure 7 (a) Risk reduction rates and (b) rejection rates, as estimated to result from the application of two-class sampling plans, with $n = 3, 5, 10, 30$, or 50 samples per lot. Each lot is sampled and the sample size is 10 g. The mean log concentrations of the PIF are assumed to be -3 log CFU/g (category C, triangles), -4 log CFU/g (category B, diamonds), or -5 log CFU/g (category A, squares). The between-lot variability (σ_b) is 0.5 (closed markers) and 0.8 (open markers). Risk reduction is relative to no sampling. Data were taken from FAO/WHO (2006).

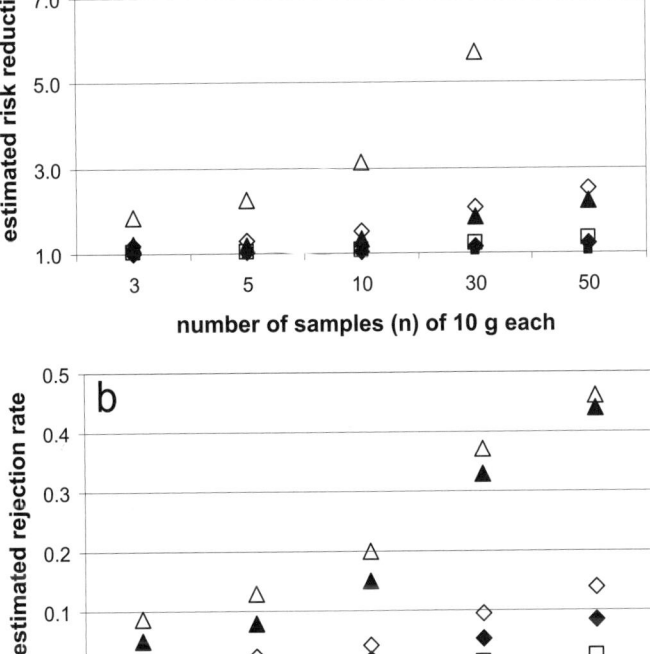

compared with the situation that no sampling is done. When 50 samples of 10 g are taken from this best category A, 0.96% ($\sigma_b = 0.5$), respectively, 2.3% ($\sigma_b = 0.8$) of the lots will be rejected, depending on the assumed between-lot standard deviation (σ_b) (Fig. 7b).

If the current mean log concentration is higher (category C; −3 log CFU/g or 1 CFU/1 kg; triangles in Fig. 7), the risk can be reduced by several units, especially if the between-lot standard deviation (σ_b) is high. This means that the batches in the "tail" of the distribution of Fig. 6 will be taken from the market. Please note that in such a case the rejection rates (Fig. 7b) mount up to 0.46. This would mean that a producer of PIF of category C could sell only little more than half of its products.

MC aim at reducing the public health burden by taking those lots out of the market that have the highest probability of causing infection. It was envisioned, however, that the application of MC would have more effects beyond the effect of taking contaminated lots from the market. Lots of PIF that do not comply with the MC cannot be sold as infant formula. Such lots have to be reworked into other products that are not intended for infants and will be sold at lower economic value. A high proportion of rejected lots, such as experienced by a category C producer in Fig. 6, implies a considerable economic loss to the producer. The rejection of lots that do not comply with the MC is seen as a strong incentive to producers to upgrade the level of hygiene during manufacturing. When a manufacturer is able to reduce the average level of *E. sakazakii* (e.g., from category C to category B), not only an increase in product safety is achieved, but also direct economic advantage is gained. This incentive will be most profound with manufacturers that are currently in the category with the highest contamination levels. From a worldwide perspective the average level of contamination will likely to be reduced. This effect generates a reduction of health risk after a certain time, which adds to the direct reduction of risk achieved by taking those lots off the market that do not meet the MC.

Here we see a very close integration of the concepts of risk assessment and the concept of setting MC as a tool to manage food safety. Moreover, the output of the risk assessment was chosen in a format that would allow risk managers to make an informed choice about the potential application of MC, albeit its uncertainties. Both the benefit for public health (risk reduction) and the implications for the producers (lot rejection) can be calculated for each type of sampling plan and weighed.

Preparation and Handling Scenarios

During the recent expert consultation, the RAModel was used to test handling scenarios for their effect on the number of *E. sakazakii*, assuming that

Case Study with *Enterobacter sakazakii* in Infant Formula   197

1 CFU would initially be present in each prepared bottle if it is contaminated (FAO/WHO, 2006). A small selection of results, indicating the effects of potential control measures, is shown below. As mentioned before, the risk is expressed as relative risk, which is the predicted number of *E. sakazakii* relative to the number predicted in the baseline scenario. In each series of scenarios one scenario was chosen and indicated as baseline. In each table and figure a different baseline scenario applies. Unless indicated otherwise, relative risks may only be compared directly within one table or figure and not across tables or figures.

Table 3 shows the results of eight different scenarios where PIF is reconstituted with water of 30°C. The ambient temperature in each scenario is 30°C representing a warm climate or nursery environment in any climate. As can be seen in the upper row, none of the handling scenarios have an increased risk compared with the baseline as long as the feeding time is short (20 min). Prolonged feeding for 2 h (lower row), however, results in an increase of the risk in three of the four scenarios, depending on the holding temperatures prior to feeding and the resulting temperature during feeding. The effect of holding time is illustrated further in Table 4, where relative risk

Table 3 Relative increase in risk of different preparation, storage, and handling practices for formula at warm ambient temperature (30°C)[a]

Feeding	Refrigeration (4 h) at 4°C after reconstitution		Storage (1 h) without refrigeration after reconstitution	
	No rewarming	Rewarming (37°C)	No rewarming	Rewarming (37°C)
Short (20 min)	1	1	1[b]	1
Long (2 h)	1	8	2.8	15

[a]PIF is reconstituted with water of 30°C, stored, rewarmed (if applicable) for 15 min, and fed at the resulting temperature for a short or prolonged period. Data were taken from FAO/WHO (2006).
[b]Baseline scenario.

Table 4 Effect of the duration of feeding on relative risk[a]

Feeding time (h)	Refrigeration (4 h) at 4°C after reconstitution	Storage (1 h) without refrigeration after reconstitution
0.33	1[b]	1
1	1	2
2	8	16
4	381	724
6	15,198	28,736

[a]PIF is reconstituted with water of 30°C, held for 4 h in a refrigerator of 4°C or for 1 h at room temperature, rewarmed to 37°C for 15 min, and fed at 37°C for the period indicated. Risk is expressed as relative risk. Data were taken from FAO/WHO (2006).
[b]Baseline scenario.

increases largely with holding time at 37°C. This effect was shown in the previous risk assessment as well (FAO/WHO, 2004).

The results suggest that prevention of prolonged holding and/or feeding times might be an effective control measure for the risk of *E. sakazakii* infection. Prolonged feeding may occur during tube feeding but also in the case the infant is fed small amounts from the same bottle over a longer period of time. Control measures may include labeling, advice to health care professionals, and also the supply of smaller syringes and bottles encouraging caregivers to prepare small quantities of formula.

The temperature of water used for reconstitution has a profound effect on the relative risk. Figure 8 shows the relative risk of a bottle of formula when refrigerated for 4 h at 4°C, rewarmed to 37°C, and then fed for an extended period of 2 h. Please note that the risk in Fig. 8 is expressed relative to the same baseline scenario as in Table 3: no refrigeration, no rewarming, and a short feeding. The relative risk varies from +2 for reconstitution with water of 10°C to +83 for water at 50°C. The use of water of 60°C leads to a small increase in relative risk, due to partial inactivation of intrinsic contamination at 60°C and some potential growth during cooling. Reconstitution at 70°C decreases the risk by a factor of >100,000 due to inactivation of *E. sakazakii*, which is a vegetative microorganism.

The dramatic risk reduction of reconstituting at 70°C suggests that such practice, in fact an on-site pasteurization step, is an effective control measure for *E. sakazakii* in PIF. It should be considered, however, that heating the PIF to such a temperature has other effects too. Many varieties of PIF have not been formulated to be used at 70°C and heat-sensitive constituents, such as vitamins, may be affected, but there are indications that the loss of vitamins is limited (FAO/WHO, 2006). Moreover, many kitchens in hospitals (not to speak of homes) are not equipped with the suitable equipment to keep the temperature at 70°C exactly, leading to either overheating or to ineffective pasteurization. Finally, hot infant formula may scald and harm both infants and caregivers when it is handled before sufficient cooling. This shows that the RAModel can give indications for control measures, but the risk manager should consider a multitude of side effects.

In many countries instructions on the label indicate that the water for reconstitution should be brought to a full boil and then cooled down until about 40 to 50°C. According to Fig. 8 and many other simulations (data not shown), these temperatures result in the greatest risks for the temperatures considered unless the formula is consumed immediately. As a control measure, reconstituting the PIF at lower temperatures (10 or 20°C) might be advised on the label. Alternatively, if the PIF has to be reconstituted at 40 to 50°C, e.g., to ensure proper homogenization of the powder, rapid cooling

Case Study with *Enterobacter sakazakii* in Infant Formula

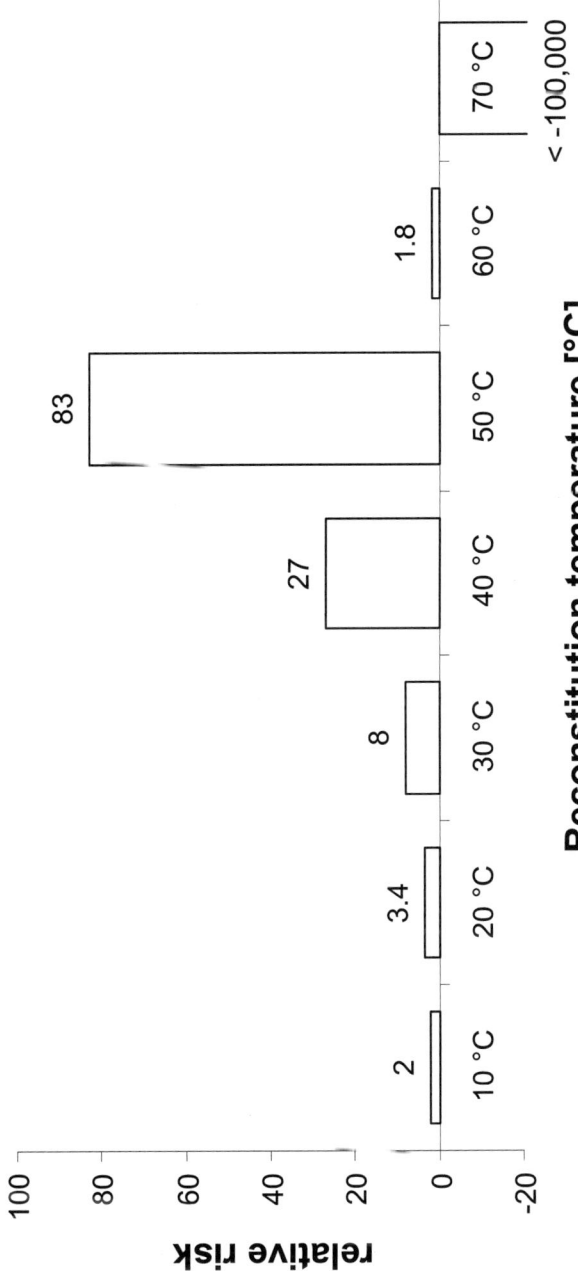

Figure 8 Relative risk of reconstituting PIF with water of various temperatures. Risk is relative to the same baseline scenario as in Table 3. Data were taken from FAO/WHO (2006).

under running water or in an ice bath might be recommended to prevent growth during cooling, holding, and feeding. One should keep in mind, however, that not in every situation is cold (and safe) water or an ice bath readily available in homes and in neonatal care institutions.

When refrigeration is available, the risk increase of holding one bottle of infant formula in a refrigerator for 4 h depends on the temperature of the refrigerator but is very limited (Fig. 9). Holding larger batches of 1 liter or even 25 liters of formula, as shown in Table 5, results in a considerable increase in relative risk even when the batch is refrigerated due to a slower cooling rate of large batches. If such practices do occur indeed, control measures targeting prevention of cooling large volumes might be useful.

The scales of relative risk resulting from various factors vary largely. Some factors have a limited effect, while others affect the relative risk by many orders of magnitude. Varying the refrigerator temperature between 2 and 10°C, for example, has an effect of 0 to 36% (relative risk 1.0 to 1.36) as seen in Fig. 9. Other practices affect the risk by several factors, such as the application of MC that may reduce the risk by a factor 1 to 7 (Fig. 7). Rewarming the formula to 37°C increases the risk by the same order of magnitude (Table 3). Some factors have a very profound impact on relative risk. The temperature of the water used for reconstitution (Fig. 8), and the duration of feeding or hangtime (Table 4) and the volume of the batch to be cooled (Table 5), may affect the risk by 100-fold to 1,000-fold. The risk assessment model will be

Figure 9 Relative risk associated with placing the bottle of formula in refrigerators with various air temperatures for 4 h, then rewarming to 37°C and feeding for an extended period of 2 h in a warm room at 30°C. Bottles are reconstituted with water of 10°C (×), 30°C (□), and 50°C (▲). Data were taken from FAO/WHO (2006). The baseline scenario for series is refrigeration at 2°C.

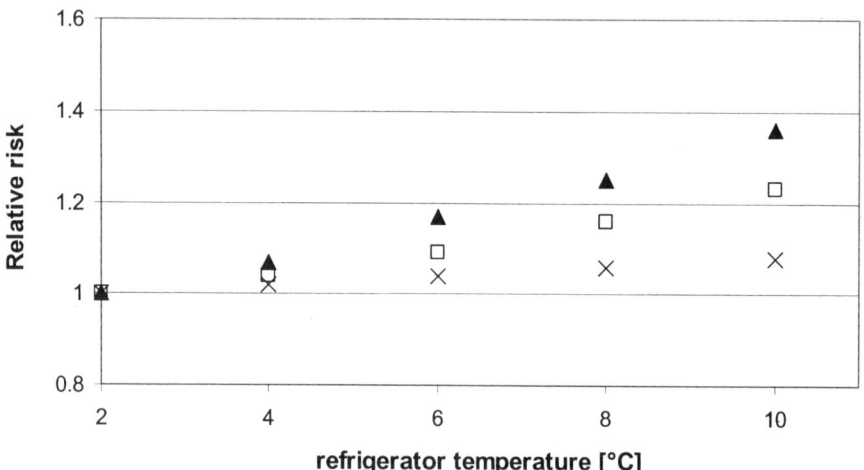

Table 5 Relative risk associated with reconstituting PIF in bottles and in containers of 1 and 25 liters[a]

Condition	Temperature of water for reconstitution	
	30°C	50°C
Mix in bottle, cool and hold for 1 h, no rewarming, and feed for 20 min	1[b]	5
Mix in 25-liter container, hold 1 h without refrigeration, feed for 30 min	1	1.6
Mix in 25-liter container for 1 h, 6 h refrigeration, rewarm, feed for 30 min	56	72,463
Mix in 1-liter container, hold 1 h without refrigeration, feed for 30 min	1	4.2
Mix in 1-liter container for 1 h, 6 h refrigeration, rewarm, feed for 30 min	1.2	45

[a]Both unrefrigerated holding and feeding are at 30°C. Data were taken from FAO/WHO (2006).
[b]Baseline scenario.

available for risk assessors and risk managers to simulate practices relevant for specific target groups and situations and to compare the effect of control measures.

RISK COMMUNICATION AND RISK MANAGEMENT

Many control measures to reduce the risk of *E. sakazakii* involve risk communication between stakeholders. And in the case of PIF there are more stakeholders than with food products that are intended for other groups in society. In a hospital situation medical doctors and other personnel decide on the type and way of feeding an infant and they often take care of the implementation. But also in the home situation professionals in infant health care do influence preparation practices by giving advice to the mother and to other caregivers. To implement risk management options that aim to change preparation and handling practices, risk communication has to target both the professionals in health care and the parents and other caregivers.

Risk communication should also be considered in the light of the nutrition that is the first choice for infants: breast milk. Mothers are encouraged to exclusively breastfeed for 6 months and health care workers should provide them with skilled support to do so. But breast milk is not always available and PIF is often the most agreeable and cost-effective alternative to provide a newborn infant with the nutrition it needs to thrive.

A potential control measure for *E. sakazakii* is to use commercially sterilized infant formula instead of powdered formula. The recent expert consultation (FAO/WHO, 2006) advised the FAO, WHO, Codex, their member countries, nongovernmental organizations, and the scientific community: "In situations where infants are not breastfed, caregivers of high-risk infants should be encouraged to use, whenever possible and appropriate, commercially sterile liquid formula or formula which has undergone an effective point-of-use decontamination procedure." But these sterile liquid products are more expensive, difficult to transport, and have so far not been available in all the varieties required for specific high-risk groups of infants.

Moreover, it should be kept in mind that sterile formula is only sterile as long as the package is unopened. There is anecdotal evidence, even from rich and industrialized countries, that sterile packages are also redistributed and stored. The reason was that a medical doctor prescribed a feeding volume of 65 ml, while the commercial package contained only 60 ml. Milk kitchen personnel thus redistributed the sterile formula once a day in other bottles and stored them until use in the refrigerator. Such practices may lead to extrinsic contamination, which is not considered in the RAModel, and may lead to increased risk.

To effectively decontaminate powdered formula at the point of use, procedures should be developed and tested to ensure that the decontamination by heat treatment or other methods will safeguard the nutritional value of the PIF and will not harm infants or caregivers due to scalding.

Producers of PIF communicate with caregivers by carefully labeling their products with instructions for preparation. But labeling alone will not be sufficient, certainly not if caregivers get other messages from other sources or if they experience problems in implementing the instructions. In The Netherlands, for example, all producers recently changed the preparation advice on the label. The option to prepare infant formula for 24 h (up to 1 liter) at once and to store it in the refrigerator is no longer mentioned. This option had been on the label of some types of PIF before. All labels currently prescribe that each portion of PIF is reconstituted separately and consumed immediately. However, stores that sell supplies for baby care offer 1-liter measuring cups on their shelves along with the baby bottles. These cups are equipped with a lid and are specifically meant to reconstitute a 24-h supply of infant formula and to store it in the refrigerator. For many parents and other caregivers these measuring cups are a very clear message that 1 liter of infant formula can be stored in the refrigerator for 24 h. As the practice of reconstituting once per day is also commonly perceived as convenient, the instructions on the label are likely to be disregarded.

Risk communication from users to risk managers is also required. To be able to weigh the various options for control measures, it is imperative that the actual preparation and handling practices become known. Risk assessors can then use the RAModel to compare the options, based on better assumptions regarding temperature, holding times, and cooling rates. Consequently risk managers may be able to design and implement control measures that will effectively reduce the risk, but at the same time take into account all the relevant factors (see Fig. 1) to make sure that the control measures will be feasible and acceptable and can be applied for those infants who are currently most at risk.

REFERENCES

Bar-Oz, B., A. Preminger, O. Peleg, C. Block, and I. Arad. 2001. *Enterobacter sakazakii* infection in the newborn. *Acta Paediatr.* **90**:356–358.

Biering, G., S. Karlsson, N. V. C. Clark, K. E. Jonsdottir, P. Ludvigsson, and O. Steingrimsson. 1989. Three cases of neonatal meningitis caused by *Enterobacter sakazakii* in powdered milk. *J. Clin. Microbiol.* **27**:2054–2056.

Block, C., O. Peleg, N. Minster, B. Bar-Oz, A. Simhon, I. Arad, and M. Shapiro. 2002. Cluster of neonatal infections in Jerusalem due to unusual biochemical variant of *Enterobacter sakazakii*. *Eur. J. Clin. Microbiol. Infect. Dis.* **21**:613–616.

Bowen, A. B., and C. R. Braden. 2006. Invasive *Enterobacter sakazakii* disease in infants. *Emerg. Infect. Dis.* **12**:1195–1189.

Breeuwer, P., A. Lardeau, M. Peterz, and H. M. Joosten. 2003. Desiccation and heat tolerance of *Enterobacter sakazakii*. *J. Appl. Microbiol.* **95**:967–973.

Centers for Disease Control and Prevention (CDC). 2002. *Enterobacter sakazakii* infections associated with the use of powdered infant formula—Tennessee 2001. *Morb. Mortal. Wkly. Rep.* **51**:298–300.

Coignard, B., and V. Vaillant. 2006. Infections à *Enterobacter sakazakii* associées à la consommation d'une préparation en poudre pour nourrissons. France, octobre à décembre 2004. Rapport d'investigation. Editions InVS, Saint-Maurice, France, 2006. [Online.] http://www.invs.sante.fr/display/?doc=publications/2006/infections_e_sakazakii/index.html.

Council for Agricultural Science and Technology (CAST). 2006. Using risk analysis to inform microbial food safety decisions. Issue Paper 31. CAST, Ames, IA.

Drudy, D., N. R. Mullane, T. Quinn, P. G. Wall, and S. Fanning. 2005. *Enterobacter sakazakii*: an emerging pathogen in powdered infant formula. *Clin. Infect. Dis.* **42**:996–1002.

Edelson-Mammel, S. G., and R. L. Buchanan. 2004. Thermal inactivation of *Enterobacter sakazakii* in rehydrated infant formula. *J. Food Prot.* **67**:60–63.

Edelson-Mammel, S. G., M. K. Porteus, and R. L. Buchanan. 2005. Survival of *Enterobacter sakazakii* in a dehydrated powdered infant powder. *J. Food Prot.* **68**:1900–1902.

Farmer III, J. J., M. A. Asbury, F. W. Hickman, D. J. Brenner, and the Enterobacteriaceae Study Group. 1980. *Enterobacter sakazakii*: a new species of "*Enterobacteriaceae*" isolated from clinical specimens. *Int. J. Syst. Bacteriol.* **30**:569–584.

Food and Agriculture Organization/World Health Organization (FAO/WHO). 2004 *Enterobacter sakazakii* and other microorganisms in powdered infant formula. Microbiological risk assessment series no. 6. Food and Agriculture Organization of the United Nations, Rome, Italy.

Food and Agriculture Organization/World Health Organization (FAO/WHO). 2006. *Enterobacter sakazakii* and *Salmonella* in powdered infant formula. Meeting Report, Food and

Agriculture Organization of the United Nations, Rome, Italy. 16–20 January 2006. Advance prepublication copy, 2 May 2006.

International Commission for Microbiological Specifications for Foods (ICMSF). 2002. *Microorganisms in Food. 7: Microbiological Testing in Food Safety Management*. Kluwer Academic/Plenum Publishers, New York, NY.

Iversen, C., M. Lane, and S. J. Forsythe. 2004. The growth profile, thermotolerance and biofilm formation of *Enterobacter sakazakii* grown in infant formula milk. *Lett. Appl. Microbiol.* 38:378–382.

Kandhai, M. C., M. W. Reij, L. G. M. Gorris, O. Guillaume-Gentil, and M. Van Schothorst. 2004. Occurrence of *Enterobacter sakazakii* in food production environments and households. *Lancet* 363:39–40.

Kandhai, M.C., M.W. Reij, C. Grognou, M. Van Schothorst, L. G. M. Gorris, and M. H. Zwietering. 2006. Effect of preculturing conditions on lag time and specific growth rate of *Enterobacter sakazakii* in reconstituted powdered infant formula. *Appl. Environ. Microbiol.* 72:2721–2729.

Lai, K. K. 2001. *Enterobacter sakazakii* infections among neonates, infants, children and adults: case reports and a review of the literature. *Medicine* 80:113–122.

Lawn, J. E., S. Cousens, J. E. Lawn, and J. Zupan. 2005. Neonatal survival. 1: 4 million neonatal deaths: When? Where? Why? *Lancet* 365:891–900.

Muytjens, H. L., H. C. Zanen, H. J. Zonderkamp, L. A. Kollee, I. K. Wachsmuth, and J. J. Farmer III. 1983. Analysis of 8 cases of neonatal meningitis and sepsis due to *Enterobacter sakazakii*. *J. Clin. Microbiol.* 18:115–120.

Nazarowec-White, M., and J. M. Farber. 1997. Incidence, survival, and growth of *Enterobacter sakazakii* in infant formula. *J. Food Prot.* 60:226–230.

Newell, M.-L. 2004. HIV transmission through breastfeeding: a review of available evidence. *World Health Organization Library Cataloguing-in-Publication Data*. [Online.] www.who.int/child-adolescent-health/publications/NUTRITION/ISBN_92_4_156271_4.htm.

Noriega, F. R., K. L. Kotloff, M. A. Martin, and R. S. Schwalbe. 1990. Nosocomial bacteremia caused by *Enterobacter sakazakii* and *Leuconostoc mesenteroides* resulting from extrinsic contamination of infant formula. *Pediatr. Infect. Dis.* 9:447–449.

Pagotto, F. J., M. Nazarowec-White, A. Bidawid, and J. M. Farber. 2003. *Enterobacter sakazakii*: infectivity and enterotoxin production in vitro and in vivo. *J. Food Prot.* 66:370–375.

Simmons, B. P., M. S. Gelfand, M. Haas, L. Metts, and J. Ferguson. 1989. *Enterobacter sakazakii* infections in neonates associated with intrinsic contamination of a powdered infant formula. *Infect. Control Hosp. Epidemiol.* 10:398–401.

Stoll, B., N. Hansen, A. Fanaroff, and A. Lemons. 2004. *Enterobacter sakazakii* is a rare cause of septicemia or meningitis in VLBW infants. *J. Pediatr.* 144:821–823.

Van Acker, J., F. de Smet, G. Muyldermans, A. Bougatef, A. Naessens, and S. Lauwers. 2001. Outbreak of necrotizing enterocolitis associated with *Enterobacter sakazakii* in powdered milk formula. *J. Clin. Microbiol.* 39:293–297.

Zwietering, M. 2005. Practical considerations on food safety objectives. *Food Control* 16:817–823.

Microbial Risk Analysis of Foods
Edited by Donald W. Schaffner
© 2008 ASM Press, Washington, D.C.

Communicating about Microbial Risks in Foods

William K. Hallman

INTRODUCTION

Risk communication is an essential component of efforts to assess, prevent, manage, and recover from incidents of microbial contamination and outbreaks of food-borne illness. As such, it should be made clear from the outset that good risk communication is *not* a substitute for poor risk assessment or management, nor should it be construed as a type of public relations designed to placate the public after an incident involving illness or contamination. Instead, it should be understood as being integral to *every* step of the process required to effectively deal with microbial risks.

Communicating about microbial risks begins with a good understanding of the essentials of microbiology, hazard analysis, and risk assessment. It is critically important to avoid passing on or contributing to the body of myth and misinformation that frequently serves as a barrier to public understanding of, and appropriate response to, the microbes that surround them.

Yet, while understanding the science is critical to the risk communication endeavor, it simply is not enough. To be successful, one also needs to understand human nature, how people perceive and respond to risks, their cultural beliefs and practices regarding food, how they think about "germs," and, of course, how to talk with people about all of these things. So, in addition to understanding microbiology, an effective risk communicator also needs to understand something about psychology, sociology, anthropology, epidemiology, communications, public health, and marketing.

Clearly, it is not possible to cover everything one needs to know about these topics in the short space of a single chapter. Indeed, many books are devoted

---

WILLIAM K. HALLMAN, Food Policy Institute, Rutgers, The State University of New Jersey, ASB III, 3 Rutgers Plaza, New Brunswick, NJ 08901-8520.

to each of these subjects. Instead, this chapter draws on these disciplines to help those involved with microbial risk assessment and management better understand some of the basics of effective risk communication, some of the unique challenges inherent in communicating about microbial risks, and how to avoid some of the most common mistakes.

COMMON GOALS FOR RISK COMMUNICATION

Understanding *why* and *with whom* you need to be communicating is the first step in effective microbial risk communication. While this might seem obvious, failing to do this is one of the most common mistakes risk communicators make. To avoid this error you must start with a good sense of what you are trying to accomplish through your communication efforts. That is, you must clarify your goals.

The National Research Council (NRC) (1989) identified education, advocacy, and fostering a partnership for decision making as the three common goals for risk communication. It is important to note that in accomplishing these goals, the assumed roles of the communicator and the audience differ. In each case, there are inherent assumptions about who has information worth sharing and who should be involved in the decision-making process. Problems arise when these expected roles are not clear to those involved or accepted by them. As such, it is worth examining the assumptions implicit in achieving the three goals identified by the NRC.

Often, the goal of risk communication is to provide education or information to put a new risk into context, or to help people appropriately prepare for or manage known risks. For example, this may be the goal following a hazard assessment, the discovery of a new or reemerging pathogen, the threat of purposeful or unintentional contamination, or the outbreak of disease. This may also be the case when information becomes available about new methods to prevent, manage, or remediate microbial contamination. The assumptions here are that the communicator has special expertise or information to share that would be useful to a particular audience, and that the audience has the ability to use what they learn to take appropriate actions. Under these circumstances, risk communication efforts are typically designed to provide useful and authoritative information to people so that they can choose what *they* believe is the right course of action to deal with a particular risk.

However, sometimes the goal of risk communication is to change existing beliefs, attitudes, or behaviors, persuading people to adopt a particular viewpoint or to take (or not take) specific actions. In this case, the presumption is that there is a "correct" set of beliefs that people should hold or that there is a clear course of action. It is also assumed that the communicator has special expertise or knowledge to determine these and that persuading people to adopt them is in their own best interests or in the interests of society. For example, this is usually the situation when training people to safely select, prepare, and store foods, explaining how to prevent accidental microbial contamination, boil water advisories, or in hygiene and health promotion campaigns. Under these circumstances, the efforts are typically designed to provide useful and authoritative information to people so that they will choose what the *communicator* believes is the right viewpoint or course of action.

At other times, the goal of risk communication is to collect or discuss information that will lead to better decisions. Such might be the case when speaking with managers or employees while conducting a hazard analysis or in designing safer work areas or procedures. It is also typically the case when talking with people to better understand their particular concerns about a risk or their experiences with it, and in efforts to enhance public participation in programs designed to prevent, manage, or remediate microbial risks. The presumption here is that while the communicator may have special knowledge and expertise, he or she does not have all of the answers. Indeed, it assumes that there are others who have important information, experience, or perspectives that should be taken into consideration when making decisions. Under these circumstances, the communication efforts are typically designed to *exchange* useful and authoritative information so that the right course of action can be determined.

Note that in such an exchange, the definition of "useful and authoritative" is not restricted to technical information, but may include a variety of perspectives, value judgments, experiences with, and opinions about the risks in question that may differ from those held by the communicator. These differences should be considered a strength of the exchange, since decisions made in isolation are rarely optimal. Moreover, such divergent viewpoints must be demonstrably taken into consideration if the resulting decisions are going to be accepted by those involved.

The Risks of Failing To Clarify Goals

Often, when risk communicators have not thought carefully about what they intend to accomplish, or about the implicit assumptions regarding the expected roles of the communicator and his or her audience, the result is a divergence between the communicator's stated and real underlying goals. Naturally, most risk communicators do not purposely try to mislead their audiences; to do so would be unethical (and largely ineffective). Yet, they often end up doing so because they lack clarity about their own goals and how they will accomplish them. As such, they are unable to make these clear to their audiences. Unfortunately, whether intended or not, this divergence can be seen by audiences as evidence of duplicity or untrustworthiness, undermining the efforts of the communicator.

The most common problem arises when the communicator's stated goal is to have a true dialogue with people, seeking their input about how to better prevent or manage a risk, while the apparent goal is to simply announce and defend decisions that have already been made. Frankly, there are few better ways to make people angry than to mislead them into thinking that their input is valued when it clearly is not.

Another common scenario is one in which the communicator says that the goal is to provide information so that people can make their own decisions. Meanwhile, it quickly becomes apparent to the audience that the goal is to *persuade* them that the communicator's particular viewpoint or action is the one that makes the most sense. In such cases, it usually takes only a few minutes for people to realize what is happening and for them to respond negatively. In fact, research on persuasion consistently shows that it is much more effective to tell people at the start that you are going to try to persuade them, and then proceed with persuasive arguments (O'Keefe, 2002a).

So, to avoid misleading people about your intent, it is important to decide whether your role is to simply provide information and let your audience decide what to do; to try to persuade your audience to change their beliefs or their behaviors; or to have a true dialogue where you are as interested in what others have to say as you are in getting your own message across. In thinking about your role, it should be evident that there are some cases where the communicator is clearly recognized by the audience as being in the best position to provide information, make decisions, or provide advice. In such cases, while people may have an opportunity to ask questions, information is seen by all of those involved as appropriately flowing primarily from the expert communicator to the audience. In situations that do not require shared decision making or a great deal of interaction between experts and their audiences, "one-way" communications such as lectures, TV and radio interviews, training programs, websites, brochures, or advertisements can be effective in getting limited amounts of information across.

In most other cases, however, "two-way" communications are much more appropriate and effective (Hance et al., 1988). Here, the roles of communicator and audience are shared among all of the participants, and there are opportunities allowing feedback about whether the needs of all the participants are being met and the information being discussed is understood by all parties. Such "two-way" communications are essential when the goal involves discussing complex or controversial information or where shared or consensual decision making is important.

The problem is that there is a tendency for many experts to see "one-way" communications as the appropriate strategy for *all* risk communication efforts. In part, this is because it is often faster and cheaper to engage in such efforts. It is also easier to reach large audiences with one-way mass media campaigns, and because the content of many of these communication efforts is static, the quality of the information shared is much easier to control. Moreover, because there is limited or no interaction with an audience in "one-way" communications, there are few opportunities for criticism of or questions about the authority of the expert. For experts who are uncomfortable speaking with audiences or who may possess limited social skills, two-way communications can be threatening. Yet, doing an effective job of sharing complex information, making difficult decisions, and persuading people to change their behaviors *requires* interacting with others. Therefore, it is important to choose the proper communication strategy based on the requirements of the specific situation rather than on the limitations of the communicator.

CHOOSING THE RIGHT COMMUNICATOR

Two important goals of communication are to be understood and to be believed (Taillard, 2000). Therefore, in choosing the right communicator, it is important to select someone who "knows the material," is confident in his or her ability to talk about it, and, through his or her character and actions, can inspire the trust and confidence of others.

This has long been understood. Around 350 BC, Aristotle wrote in his primer on persuasive speaking, *Rhetoric*, that while a speaker must obviously know what he is talking about, "It adds much to an orator's influence that his own character should look right." He argued that possessing a character that "looked right" depended fundamentally on the communicator's demonstrable good sense, goodwill, and good moral character (Barnes, 1984).

Since that time, much has been written about trustworthiness and credibility as being essential characteristics of good risk communicators. Hence, many scholars have tried to identify the key components of perceptions of trust and credibility, and an extensive body of literature is devoted to the issue (see, for example, Coleman, 1990; Covello, 1992, 1993; Frewer et al., 1995; Hardin, 2002; Johnson, 1999; Kasperson, 1986; Kasperson et al., 1992; Lang and Hallman, 2005; Peters et al., 1997; Poortinga and Pidgeon, 2003; Renn and Levine, 1991; Rotter, 1971). While extensive debate remains concerning the number and nature of the determinants of perceived trust and credibility, Peters et al. (1997) note that most theoretical frameworks include three basic components. If you want people to believe what you have to say and trust you as a valued partner, both the messages and the messenger must demonstrate evidence of knowledge and expertise, openness and honesty, and concern and care.

Fortunately, locating people with the technical background necessary to demonstrate evidence of knowledge and expertise around common microbial risk issues is not terribly difficult. Moreover, external evidence of this expertise, in the form of advanced degrees; published books, papers, and reports; honors and awards; and years of experience, is also available to help select appropriate experts in cases requiring specialized knowledge.

Yet, it is important to recognize that although demonstrable knowledge and expertise are critical to perceived trust and credibility, it is only one-third of the equation. The remainder points to the value of possessing and demonstrating a high degree of social and emotional intelligence (Goleman, 1995). In fact, demonstrating a high degree of knowledge and expertise in the absence of the ability to connect with ordinary individuals may reduce the persuasiveness of a communicator by emphasizing the detachment of experts from "normal" people (McGinnies and Ward, 1980; Frewer et al., 1997).

Successful communication depends largely on the relationships that develop among those involved. This is particularly true in risk communication situations, where the threats of contamination or illness can lead to strong emotional reactions, including worry, fear, frustration, and anger (Gray and Ropeik, 2002). Therefore, it is equally important to find communicators who have good social skills; who are able to talk with people easily and honestly; and who understand their concerns, can genuinely empathize with them, and are able to respond appropriately. The key here is that good communicators do not necessarily want to do all of the talking. They want to *listen* to people and help them with their problems. They adapt their approaches to best meet the needs of those with whom they are trying to communicate. They are willing to admit when they do not know the answer to a question, but will also work hard to find that answer if it exists. They do not take themselves too seriously. They care.

The problem, of course, is that proof of good social skills, honesty, empathy, caring, openness, compassion, or a sense of humor does not necessarily appear on a standard résumé or curriculum vitae. However, vetting a communicator's ability to connect with other people is as important as checking his or her other credentials. Teaching awards, evidence of volunteer work, positions of leadership in, or awards from, service, religious, or civic organizations may provide indirect indications of emotional intelligence. However, perhaps the final test of whether a person possesses the social skills necessary to make a good communicator is whether he or she is someone people would want to talk with at a party or other social gathering. If the person you are considering as lead risk communicator typically stands alone at such functions, or has difficulty making conversation about anything unrelated to his or her work, that is a bad sign. Given the importance of the threat of microbial contamination and life-threatening food-borne illnesses, it makes little sense to choose a communicator to speak in life-or-death situations whom people would normally avoid in social situations.

Unfortunately, it can be difficult to find a single person who possesses both the requisite technical expertise and the essential social skills required to serve as an effective risk communicator. Yet, messages about microbial risks constructed in the absence of appropriate technical knowledge are likely to be incorrect or misleading, while messages constructed or delivered without sensitivity to people's needs and emotions are likely to be misunderstood or mistrusted. Therefore, it may be necessary to assemble teams of people who have the necessary skills. Sometimes this means choosing a lead person with good social skills backed by a group of technical experts.

UNDERSTANDING YOUR AUDIENCE

While it is critical to clarify what is to be accomplished, and who will do the communicating, it is also important to consider with whom you need to communicate. These audiences may include professionals from a variety of disciplines, managers, employees, trainees, students, union officials, public health officials, consumers, and newspaper and television reporters, among others.

Keep in mind that it is likely that you will have multiple audiences. Marketing and advertising professionals learned long ago that there is no such thing as *"the* public," and that to be successful, they need to reach *multiple* publics. Marketers have become increasingly sophisticated in their abilities to identify and reach specific market segments with messages that predictably resonate with that particular group (Wedel and Kamakura, 1999). To be effective, risk communicators must adopt similar market segmentation approaches. For example, people's concerns and behaviors regarding microbial risks typically differ by gender, ethnicity, age, education, and income (Patil et al., 2004). Moreover, the kinds of information needed or desired by consumers, hobby chefs, commercial food handlers, food service managers, nutritionists, teachers, trainers, regulators, and other audiences are likely to differ and will require different approaches to reach them. Importantly, there are also particular segments of the population that may face severe health consequences or death from exposure to food or waterborne disease, especially pregnant women, infants, young children, the elderly, and the immunocompromised (Kendall et al., 2003). To help avoid these consequences, special efforts may be required to reach these populations with targeted messages. Again, this may seem obvious, but many risk communication efforts treat all audiences as if they were the same, attempting to meet the needs of all with a single brochure, fact sheet, poster, training session, or public meeting. It is not surprising that these efforts usually fail; different approaches and different messages are needed for different audiences.

It is also important to understand that even *within* these various audiences, people will probably differ in terms of their interest in what you have to say and how it fits with their own needs, concerns, and responsibilities. They may also come from different cultural backgrounds and may therefore have differing perspectives on the nature and acceptability of particular microbial risks, and perhaps differing views on the proper ways to prevent and treat food-borne illnesses in particular. They will also likely have differing levels of experience, education, and literacy, and so may possess greater or lesser abilities to understand, put into context, or put into practice the information or ideas you have to offer. They may also have varying abilities to express or clarify their own ideas, questions, or needs.

According to 2000 U.S. Census figures, nearly 18% of the population speaks a language other than English at home, with more than 21 million people reporting speaking English with some difficulty (8.1% of the population), including about 3.4 million Americans reporting speaking English "not at all" (1.3% of the U.S. population). About 4.7% of the population lives in a "linguistically isolated household" where no member over the age of 13 speaks English "very well." To be effective, risk communication efforts may be required in multiple languages in addition to English. Indeed, failure to translate risk communication materials into commonly spoken languages might be construed by those who do not speak English as evidence that they are considered "unequal," not worth communicating with, or not in a group affected by the risk. Failure to take appropriate actions to communicate about risks with workers who speak languages other than English might also raise civil rights issues or other legal liabilities. While Spanish is the secondary language most often spoken in the United States (U.S. Census, 2000), the National Virtual Translation Center (2006) estimates that 176 languages are spoken in the United States and the Census Bureau tracks the locations and numbers of speakers of 30 common languages and three groups of less commonly spoken languages in the United States. Note that some languages may lack the vocabulary to describe the microbial world. This may be important in particular when developing risk communication materials for the developing world.

While many have difficulty understanding spoken English, the U.S. Department of Education (2003) also estimates that approximately 30 million American adults (14% of the adult population) have below basic English prose literacy levels, meaning that they possess "no more than the most *simple* and *concrete* literacy skills." More than half (55%) of those who did not finish high school fall into this category, as do 44% of those who spoke no English before starting school, 39% of all Hispanic adults, 20% of all black adults, 26% of those older than 65, and 21% of adults with multiple disabilities. An additional 63 million American adults (29%) have only basic skills, suggesting that they can perform only simple, everyday literacy activities. So, it is important to remember than many within your audiences may be unable to read or write, making written instructions, warnings, or operating procedures incomprehensible.

In response, collections of symbols have been created to quickly communicate warning information without requiring literacy or as a supplement to text-only labels. These symbols range from the familiar "skull and crossbones" used to indicate toxic materials to newer warning symbols developed by the American National Standards Institute (ANSI) and the International Organization for Standardization (ISO) covering a large variety of hazards. However, many symbols currently in use are poorly understood (Davies et al., 1998), especially among the elderly (Lesch, 2003), and unfortunately, the meanings of "universal" warning icons may differ across cultures (Rother, 2006).

It should also be understood that many people, including the literate, are relatively innumerate. They have difficulty grasping the magnitudes of very large and very small numbers and have a hard time interpreting the meanings of fractions, proportions, and probabilities (Paulos, 1988). As such, communications involving mathematical operations or statistical descriptions may not be easily understood. The English language also offers an extensive assortment of expressions to describe quantities and frequencies, most of which are vague or easily interpreted in a variety of ways. Thus, the meaning of words such as *often, a few, little, many, frequently,* and other descriptors can have different meanings for different people (Moxley and Sanford, 1993).

Because of all these differences among audiences and the potential barriers to communication associated with them, it is imperative to tailor approaches, messages, and channels for communicating so that they are most appropriate to the needs, desires, and abilities of a particular audience. Creating general materials designed to meet the needs of all audiences risks meeting the needs of none.

Finding the Right Starting Points To Communicate with Your Audience

Once the goals for communication have been clarified, a risk communicator must determine where to begin. Given that audiences and the reasons for communicating differ, the appropriate starting points for communications will also vary. Nevertheless, whether the goal is to inform people, persuade people, or have a conversation with people, one should begin with an assessment of what people know, how they know it, and what they want to know about now.

The big mistake is to use an expert model of what people *need* to know (as opposed to what they *want* to know) as the starting point for *all* microbial risk communications. Typically, this strategy begins by asking a group of experts to reach some consensus about the information that lay people need to know to understand the science behind a risk. The second step is usually to try to translate the science into simple language and graphics so that nonexperts can understand it (so-called "dumbing down" the science). The last step is usually to create a set of communications materials to be disseminated to as many people as possible.

While it is often useful to help people understand the science behind microbial risks, communication strategies based solely on an expert model are usually unsuccessful. One reason is that experts tend to overestimate what ordinary people know about science. So, for example, many educational efforts assume that laypeople have a level of understanding of basic biology that, in general, is not borne out by surveys of the public (e.g., National Science Board, 2006). As such, some of these educational materials miss the mark simply because many people are not able to put the information provided into any meaningful context. These efforts fail because they try to build on a foundation that simply does not exist.

Some risk communication efforts based on the expert model fail because, understanding that many people lack the necessary scientific foundation to understand more complex ideas and information, a communicator insists on trying to build such a foundation, brick by tedious brick. These attempts are unsuccessful because, while most people *say* they are interested in science (National Science Board, 2006), experts often overestimate the interest that lay people have in learning the *details* of science. This miscalculation is understandable, since it is rather discouraging for one to think that not everyone is absolutely fascinated by the scientific minutiae that form the basis of an expert's intellectual pursuits. Yet, it is this firm belief that everyone *is* interested in the specifics of their work that explains why so many experts dine alone (see "Choosing the right communicator" above).

Frankly, outside of formal education, most laypeople lack the time and tolerance for tedious talks about scientific topics. Instead, people are much more willing to pay attention to information that answers *their* questions or meets their specific needs. Because of this, in some situations, efforts to "educate" the public might be more effective if their usual formats were inverted. That is, while it may seem necessary to spend significant time in laying the scientific groundwork to properly answer people's questions, it is unlikely that people will pay attention long enough to ultimately have their questions answered. Instead, people are more likely to sit still for an approach that attempts to answer common questions first, but then backs up those answers with the scientific details that permit a fuller explanation.

Most often, however, risk communication efforts based on an expert model fail because they are based on a faulty premise. The expert or "knowledge deficit" model (Einsiedel, 2000; Hansen et al., 2003; Wynne and Irwin, 1996) assumes that "if people just knew the facts" they would "reach the right conclusions" or "make the right decisions." Unfortunately, the data show that this is rarely the case (Weinstein, 1988). The ineffectiveness of simply providing information is also regularly confirmed in our daily lives. If "simply providing the facts" were really effective in convincing people to adopt a particular point of view or to change their behaviors, there would be no need to read this chapter. There would also be no need for politics, advertising, or rhetoric. If such a strategy were truly effective, people would no longer smoke, eat or drink to excess, or practice unsafe sex. There would also be no more wars, we would all drive the same model car, everyone would wash their hands after using the toilet, and every child would eat peas.

Although we clearly do not live in such a utopian society, it is easy to see why the knowledge deficit model has such an appeal to scientists—it assumes that most people are just like them, only perhaps not as smart or well informed. Part of the knowledge deficit model assumes that, "in assessing the risks presented by food, experts and consumers are, or should be, doing the same thing. The idea is that experts do this thing well and consumers do not" (Hansen et al., 2003). It also assumes that the experts are "right" and that laypeople are "wrong" because nonexperts lack the requisite scientific knowledge and understanding of the issues to make proper decisions. Therefore, the obvious solution to this deficit in knowledge is to educate people, to remediate their shortcomings, and to encourage them to adopt the "informed" views of the experts.

However, the evidence suggests that this strategy is usually unsuccessful. While consumers *do* use scientific information in their decisions about risk, unlike formal risk assessors, it is not the *only* information they use when considering a risk. So, while scientific knowledge does seem to have an impact on attitudes, the relationship is often modest and nonlinear (Sturgis and Allum, 2004).

Scholars have suggested several reasons for this. The impacts of scientific knowledge may be moderated or "contextualized" by other types of knowledge that may include an understanding of how science and scientists come to know things (the methods and processes of science) and how science is funded, controlled, and organized (an issue of trust) (Wynne, 1992). How scientific information is incorporated into how people think about risks can also be strongly influenced by cultural norms (Douglas and Wildavsky, 1982) or by how consistent they are with particular worldviews, core beliefs, or values held by individuals (Slovic and Peters, 1998). Moreover, an important component to the influence of scientific knowledge is the extent to which it seems to integrate and be consistent with personal experience (Jasanoff, 2000). Finally, people may have concerns about risks other than those that are typically addressed by science (Hansen et al., 2003). Therefore, as Peters (2000) suggests, there are many reasons that consumers may disagree with or fail to follow the science-based recommendations made by experts that have little to do with a lack of understanding of the science.

This does not mean that we should abandon efforts to help people understand the science behind microbial risks. We just need to be realistic about how effective such efforts are likely to be by themselves. In cases where experts and laypeople clearly perceive each other as having common goals and concerns, the dissemination of scientific information and expert advice may be appropriate, appreciated, and exactly what is called for. The mistake is in overestimating the extent to which experts and laypeople share common goals, values, and concerns, and in applying the knowledge deficit model to situations where this commonality does not exist. In such situations it is clear that simply "translating science for the common man" is not an effective risk communication strategy.

WHAT DO PEOPLE KNOW ABOUT MICROBIOLOGY?

Whether the goal of a communication effort involves specifically talking with people about the science of microbial risks, it *is* important to understand what people know (or think they know) about microbiology so that messages can be designed that connect with people's underlying mental models of the microbial world (Morgan et al., 2002). Because it fits with what they already know, such information is much more likely to be remembered and incorporated into how people think about how microbes work (Petty and Cacioppo, 1986). Since information that is consistent with or easily integrated into existing viewpoints can reinforce what people already believe, it is useful to understand both what people generally have right and wrong about the microbial world.

First, most Americans say they are familiar with microbes such as bacteria and viruses. In a national survey of American's knowledge and opinions about basic microbiological concepts, Hallman and Condry (2005) found that only 1 in 4 said that they know "little" or "nothing at all" about "germs" such as bacteria and viruses. The remaining three-quarters said that they knew "some" (21%), "a moderate amount" (36%), or "a lot" (22%) about "germs." Moreover, 92% agreed that they knew enough about germs to keep themselves healthy.

Most people seem to know some facts about bacteria and viruses. For example, Hallman and Condry (2005) found that nearly every American (97%) knows "that there are always germs on the human body," and 91% say that that "harmful germs are virtually everywhere." Nearly every American (96%) also knows that "people can spread a virus to others even if they have no symptoms themselves," and 87% believe that "a virus can live in a person without causing an illness."

While acknowledging their ubiquitous presence, Americans seem to be unsure whether exposure to germs is a positive or negative phenomenon. For example, most Americans (84%) believe that "some exposure to germs is a good thing," and 7 in 10 think that "children who have little exposure to germs often have more severe illnesses later on." At the same time, three-quarters agree that "to stay healthy, it is important to avoid as many germs as possible" and 92% believe that "to stay healthy, it is important to keep your home as clean as possible."

While avoiding germs and keeping one's home clean are vigorously endorsed by most Americans as a way to protect health, 94% also acknowledge that bacteria and viruses cannot be seen by the naked eye. As such, people use other visible indicators as a sign that bacteria and viruses are present. For example, 86% believe that "mold" is a good indication that bacteria or viruses are also present. In addition, about three-quarters believe that both "dirt or filth" and "bad smells" are a good indication of the presence of bacteria or viruses and 45% believe that "dust" is a good indicator. Thus, microbial risk communicators should bear in mind that people generally rely on visual or olfactory clues about the presence or absence of microbes. The absence of such situational indicators may give people false confidence that there is little risk of microbial contamination. Many people simply believe that if hands, surfaces, fabrics, utensils, and equipment look "clean" it means that they are free of microbial contamination. Yet, Sharp and Walker's (2003) evaluation of the hygiene of communal kitchens clearly shows that visual hygiene assessments are not a good indicator of microbial contamination.

Beliefs about the Ecology of Microbes

Americans hold many widespread misconceptions about the biology of bacteria and viruses. For example, in their survey, Hallman and Condry (2005) found that many Americans seem unaware of the existence of anaerobic bacteria; only one-third (37%) of Americans know that not all bacteria require oxygen to live. Even fewer (28%) understand that viruses cannot reproduce outside a living organism. Many also seem to have an anthropomorphized view of germs or see them as animal-like "predators." About one-quarter believe that "germs can sense when people are nearby," nearly one-third believe that "germs can sense which people are most vulnerable," and nearly two-thirds believe that "germs move to places that make it easier for them to infect people."

Beliefs about Where Germs Are
It is also important for risk communicators to understand laypeople's expectations about the likelihood of the presence of germs in particular locations and situations. Hallman and Condry (2005) asked a random sample of 275 Americans to indicate the relative number of germs they thought could be found in nine different locations. Using a scale of 0 "no germs" to 10 "extreme number of germs," the respondents appropriately indicated that wet surfaces (mean = 6.7, SD 2.4) typically had more germs than dry surfaces (mean = 5.1, SD 2.6). However, they also indicated that public places/surfaces such as public toilet seats (mean = 8.4, SD 2.2), schools (mean = 8.2, SD 2.1), door knobs (mean = 7.8, SD 2.3), and hospitals (mean = 7.6, SD 2.6) were likely to have more germs than private places/surfaces, including their own kitchen sink (mean = 6.6, SD 2.7), the floor in their house (mean = 5.7, SD 2.6), or the toilet seat in their home (mean = 5.6, SD 2.7).

In terms of the body, respondents reported (using the same 0 to 10 scale), that germs were most numerous in feces (mean = 8.7, SD 2.0), nasal discharge (mean = 8.1, SD 2.2), saliva (mean = 7.6, SD 2.4), and blood (mean = 6.5, SD 3.0). They indicated that germs were less numerous in hair (mean = 5.2, SD 2.7) and in tears (mean = 3.9, SD 2.8).

Beliefs about Germs and Illness
Hallman and Condry (2005) found that most Americans believe that exposure to germs does not always lead to illness; only about one in ten believes that "once a germ gets inside your body, you will always get sick." Most Americans also understand that that there is typically a latency period between exposure and illness, but about one in five believes that "most germs make people sick as soon as they are inside the body."

However, laypeople have varying notions about the nature of dose-response relationships when it comes to illnesses caused by bacteria and viruses. For example, about one in three Americans believes that "for most germs, any amount of exposure will make you sick," while more than one-third (38%) believe that "the severity of illness depends on the number of germs you are exposed to."

Beliefs about Killing Germs
In its 2006 report of science literacy, the National Science Board notes that 54% of U.S. and 46% of European respondents recognized that the statement "Antibiotics kill viruses as well as bacteria" is false. The next highest percentage of correct responses was in South Korea (30%), followed by Japan (23%), Malaysia (21%), Russia (18%), and China (18%).

Hallman and Condry (2005) asked 262 Americans to rate a dozen agents according to their effectiveness at killing germs. On a scale of 0 "not at all effective" and 10 "completely effective," the respondents ranked bleach as most effective (mean = 8.4, SD 2.0) at killing germs. They ranked alcohol (mean = 7.3, SD 2.5), Lysol (mean = 7.2, SD 2.2), and ammonia (mean = 7.0, SD 2.5) as equally effective at the second tier of germ killing. Mouthwash (mean = 6.6, SD 2.2) and antibacterial soap (mean = 6.5, SD 2.4) were judged more effective at eliminating germs than either conventional hand soap (mean = 5.6, SD 2.5) or laundry detergent (mean = 5.1, SD 2.5). Sunlight was judged as only somewhat effective (mean = 4.7, SD 2.8) as a germ fighter, but was rated more than twice as effective at eradicating germs as aspirin (mean = 2.2, SD 2.8) and nearly three times more effective than either tap water (mean = 1.7, SD 2.1) or air freshener (mean = 1.7, SD 2.2).

Of particular interest is the fact that more than half of American consumers (55%) believe that antibacterial soaps are more effective at killing germs than conventional soaps (Hallman and Condry, 2005). However, Sharp et al. (2001) have shown that while antibacterial soaps containing triclosan are more effective than conventional liquid and bar soaps at killing germs in vitro, when the three soaps were compared for their ability to reduce microbial counts on the hands, no differences were observed between the three products.

HOW DO PEOPLE COME TO KNOW ABOUT MICROBES?

People typically learn about "germs" in formal ways during their school years, in classes focused on biology, health, or occasionally in what used to be called "home economics." However, unless their jobs require it, once they graduate, there is little formal opportunity for most people to learn more about the microbial world. As such, most laypeople continue to learn about microbes from informal sources including friends, family, and the popular media.

Nancy Tomes (2000) suggests that what consumers do know about bacteria and viruses comes from decades of advertisements for products such as toothpaste, mouthwash, and household cleaners and that this has been the case since "germ theory" became generally accepted. Thus, this "received knowledge," gleaned from more than a century of advertising, has become part of the culture, gaining credence as parents in each succeeding generation inculcated their own children with the lessons from these advertisements, though many of the campaigns and some of the products they promoted have now been long forgotten.

The Legacy of Advertising "Germs"

Perhaps the best remembered and most influential advertising concerning "germs" began with Listerine antiseptic. Invented in 1879, it was named after the English physician Sir Joseph Lister, who had performed the first antiseptic surgery in 1865 and was originally marketed by the Lambert Company only to medical professionals as a disinfectant for surgical procedures (Pfizer, 2005). It was then sold as a multipurpose antiseptic "for both internal and external use," and was recommended both for treating gonorrhea (Young, 1992) and as a floor cleaner (Levitt and Dubner, 2005).

In 1895, the Lambert Company marketed Listerine to the dental profession as a powerful oral antiseptic. By 1914, Listerine became one of the first prescription products to be available over the counter (Pfizer, 2005a). However, the product did not achieve real success until it was heavily marketed as a "mouthwash." Beginning in the 1920s, the marketing of Listerine featured a sustained advertising campaign depicting the unhappiness wrought by "halitosis," a scientific-sounding term invented to describe bad breath supposedly caused by bacteria growing unchecked in the mouth (Marchand, 1985). Best known for ad copy explaining why an attractive woman was "often a bridesmaid, never a bride," why the qualified job hunter had "two strikes against him," and what a doctor "whose practice includes hundreds of the better class" knows about *nice* women, the campaign traded on the social stigma associated with mouth odors (Lambert Pharmacal Company, 1956, 1946, 1929a). However, while many advertisements "educated" consumers about the germs that caused "bad breath," Listerine advertisements also featured copy explaining the efficacy of the product's antiseptic action against a variety of threats, including ringworm, "street car colds," and even dandruff (Lambert Pharmacal Company, 1917, 1932, 1929b, 1937).

As part of their marketing campaigns, the Listerine advertisements also "educated" consumers, explaining the action of germs, including how "wet feet, sudden changes of temperature, and overexertion weaken body resistance so that germs in the mouth get the upper hand." However, in one advertisement they also reported that scientists suspected that filterable viruses might be the cause of the common cold (Lambert Pharmacal Company, 1930a, 1934). Advertisements also warned consumers of the ubiquity of germs, explaining that hands have germs "breeding on them by the millions," "that wherever children gather . . . you're sure to find germs," and perhaps more menacingly, "No matter where you go now . . . you can't escape from germs!" (Lambert Pharmacal Company, 1930b, 1959, 1958). In addition to alerting consumers to the pervasiveness of germs, ads for Listerine made them visible, featuring sketches of microscopes and the contents of petri dishes. The ads also gave germs names, mentioning specific bacilli including "streptococcus hemolyticus," "staphylococcus aureus," and "bacillus typhosus" (Tomes, 1998).

Not surprisingly, the ads also emphasized the rapid and extensive germ-killing power of its product. In its advertising copy, Listerine emphasized that its products killed "200,000,000 germs in 15 seconds." Yet in a piece titled "Listerine and Other Mouth Washes," the editors of the *Journal of the American Medical Association* pointed out that, "the argument that Listerine kills 200,000,000 germs in 15 seconds means absolutely nothing. Germs are mighty small, and a bathtub full of undiluted Listerine will no doubt kill as many germs as can be gotten into the bathtub, provided the mixture is made thoroughly" (JAMA, 1931).

During this time, advertisements for other products also "educated" consumers about germs, emphasizing how their own actions (and purchases) might keep them safe. For example, an ad for the germ-killing tablets called Formamint purporting to "disinfect the most secluded corners of the throat" also included a "microphoto showing *dust germs* such as invade the throat" (A. Wulfing & Company, undated).

Advertisements for Lifebuoy, a carbolic soap, traded both on preventing the stigma of B.O.—body odor (a phrase the company coined), and its ability to remove germs. It offered free "school-sized" cakes of soap and "wash-up charts" to encourage children to wash their hands (Lever Brothers Ltd., 1938, 1939).

A variety of advertisements for Mercurochrome (Hynson, Westcott, and Dunning, Inc., 1946) and Band-Aids (Johnson and Johnson, 1943) warned Americans to carefully avoid infections by treating every scratch and cut to protect against "germs, dirt, trouble."

Ads for "sneeze proof" facial tissues also taught consumers how sneezes could "spray thousands of cold germs . . . into the air." They also anthropomorphized bacteria and viruses by adding, "they stay alive for almost an hour . . . just waiting to give a cold to father, mother, little sister" (Scott Paper Company, 1959).

Companies selling household products also traded on worries about germs. The Drackett Company (1949), makers of Drano, warned that in every sink drain there is "muck, crawling with sewer germs." Advertisements also reminded consumers of the dangers of insect and rodent vectors. Stearns' Electric Rat & Roach Paste simply claimed to "help defend against disease by killing rats, mice and roaches" (Stearns, 1942). Other advertisements were more graphic. In one such piece, the Michigan Chemical Corporation (1950), makers of a product called the Pestmaster Aerosol Insect Bomb, showed a fly perched on an infant's nose. The copy warned that "Flies are a serious problem. Cleverly disguised as a mere nuisance, flies breed in filth, indifferent as to whom or what they touch. This sticky, hairy creature is a germ-laden middleman, between contamination, disease—and you. Indelibly identified with dirt, his presence may even foreshadow tragedy. You can protect your own home and family from these carriers of dirt and disease by giving flies the only welcome they deserve: Pestmaster—and sudden death." Helpfully, the ad copy also added, "Of course, you won't spray the baby."

The Du Pont Company used similar strategies to promote the use of cellophane. According to Tomes (1998), the focal point of the advertising campaign was germ protection, especially against those germs spread by insects and other humans. Sales materials emphasized that women wanted goods untouched by others. In her book, *The Gospel of Germs*, Tomes quotes one cellophane advertisement that drives this point home: "Strange Hands. Inquisitive Hands. Dirty Hands. Touching, feeling, examining the things you buy in stores. Your sure protection against *hands-across-the-counter* touch is clear, germ-proof cellophane." So effective was this line of advertising that Tomes credits Du Pont's campaign for cellophane for ultimately creating significant consumer demand and conditioned expectations for disposable hygienic packaging and utensils, all designed to help consumers avoid other people's germs.

Contemporary Advertisements (the 99.9% Solution)

Contemporary advertisements continue to appeal to consumers' fear of germs and their consequences. They also continue to "teach" a new generation of consumers about germs using many of the devices common to earlier advertisements, including making germs visible (through photographs or illustrations), calling attention to their ubiquity, giving them names, anthropomorphizing them, and emphasizing the importance of the rapid and (nearly) complete elimination of the threat. In this world, the only good germ is a dead germ.

For example, the makers of Crest Pro-Health Oral Rinse claim that "Like the leading mouthwash it kills 99% of common germs that cause plaque, gingivitis, and bad breath" (Procter and Gamble, 2006). Interestingly, the leading mouthwash (Listerine antiseptic) only claims that it is "clinically proven to kill the germs that cause plaque, gingivitis, and bad breath" and that it "kills germs by the millions on contact" (Warner-Lambert, 2005).

The makers of Dial Complete antibacterial soap claim that it is "Over 10× more effective at killing disease-causing germs than ordinary liquid hand soaps. Is proven to kill germs such as staph, strep, *Salmonella* and *E. coli*. Is antibacterial AND antimicrobial—kills bacteria and certain strains of yeast. As effective at killing germs as hand soaps used in hospitals" (The Dial Corporation, 2005).

The makers of products such as Clorox and Lysol disinfectant sprays and wipes make claims and counterclaims about their efficacy (Bittar, 1999). A relatively new product on the market, the makers of Lysol disinfectant wipes tell consumers, "Sponges and towels do only half the job because they clean but can't kill germs. Lysol sanitizing wipes not only clean but also kill 99.9% of household germs in 30 seconds" (Reckitt Benckiser, 2005). On the same website, Reckitt Benckiser features an animated graphic showing how a single bacteria cell, dividing every 20 minutes, can become more than 8 million cells within 24 hours. It also offers a downloadable poster to teach children the importance of washing their hands.

Not to be outdone, an advertisement for Clorox disinfecting wipes explains how easy it is for children to spread virus to others in the home: "According to a University of Arizona study, it happens this easily: A child at school touches an infected surface, touches his eyes, nose or mouth, then carries the flu virus home and spreads it to nearly 60 percent of household surfaces. Before long, your home is a hotbed of germs." It urges consumers to "get in the habit of disinfecting frequently touched surfaces." Of course, it also claims that "Clorox disinfecting wipes are easy to use and can kill 99.9 percent of the germs that cause colds and flu" (Clorox, 2005a). The Clorox.com website also features the chance to learn about germs, letting visitors know that most children catch approximately 8 colds a year, that viruses can live for 72 hours on common classroom surfaces, that 80% of germs are spread by touching surfaces, that students can touch 300 surfaces in a half hour, that the average number of bacteria on a desk surface "where you rest your hand" is 10 million, and naturally, that 99.9% is the "amount of cold and flu germs that can be eliminated by disinfecting once a day" (Clorox, 2005b).

Kimberly Clark (2005) has recently introduced Kleenex Anti-Viral tissues, advertising that they "trap and kill 99.9% of cold and flu viruses in the tissue," stating that they are "Virucidal against: Rhinoviruses Type 1A and 2; Influenza A and B, Respiratory Syncytial Virus (RSV) in the tissue within 15 minutes."

The SC Johnson Company (2005), makers of a line of insect control products, has introduced "Raid with GermFighter Ant & Roach Killer, the only spray that kills ants, roaches, *and* the germs they spread. It kills bugs fast—on contact—and it keeps on killing ants and roaches with residual action even after you spray, for up to four weeks. It also kills 99.9% of the household germs that household insects may leave behind. Specifically,

Raid with GermFighter Ant & Roach Killer reduces *Staphylococcus aureus* (Staph) and *Klebsiella pneumoniae* on hard, nonporous, non-food-contact surfaces."

In addition to the pervasive messages about germs present in advertising, Tomes (2000, 2002) argues that depictions of germs in the popular media and renewed anxieties about bioterrorism have also created popular narratives of disease and risk that become absorbed into the culture. Books and movies featuring "killer germs" and "superbugs" have proliferated over the past two decades, creating a kind of "antiseptic-consciousness." Glassner (1999) and Moeller (1999) argue that shocking depictions of rare germs and diseases in the popular media have made Americans excessively worried about exposure to the Ebola virus, anthrax, and flesh-eating bacteria, while they pay scant attention to microbes they are much more likely to encounter.

This fear of germs has created a market for hundreds of products incorporating antibacterial agents, including a wide variety of soaps; kitchen, window, and "all purpose" cleaners; cutting boards; clothing and carpets; house furnishings; utensils; and ball-point pens, toothbrushes, tissues, and toys, all with little evidence of their efficacy in preventing disease (Hunter, 2000). It may also account for the popularity of "waterless hand sanitizers" such as Purell, which claims that it "kills 99.99% of most common germs that may make you and your family sick" (Pfizer, 2005b).

For communicators interested in teaching people about microbial risks, it is important to remember that people's mental models of the ecology of molds, bacteria, and viruses are likely to be strongly influenced by advertisements for products designed to eliminate them. The power of advertisers to introduce and reinforce beliefs about microbes within popular culture is likely to far outstrip that of most public information campaigns. Therefore, risk communicators need to be aware of how microbial risks are represented within popular culture and must take into consideration how their own messages are likely to fit or clash with the "wisdom" conveyed as part of commercial promotions.

PERCEPTIONS OF RISK

To be effective, risk communicators need to understand something about the psychology of risk perceptions. To begin with, a great deal has been written about the psychometric bases of perceived risk (e.g., see Slovic, 2000). The psychometric paradigm uses multivariate analyses of ratings of the qualitative characteristics of hazards to create quantitative "cognitive maps" of risk attitudes and perceptions (Fischhoff et al., 1978; Slovic et al., 1986). Many of the qualitative risk characteristics have been found to be correlated with each other so factor analyses of the ratings can condense these to a smaller set of higher-order characteristics.

Factor analyses conducted on judgments of a large and diverse set of hazards have been replicated across groups of experts and laypeople in many countries. In general, these analyses have yielded two stable factors that Slovic et al. (1980) have labeled as characteristic of "unknown risks" and "dread risk." Unknown risks are those judged as new, unknown to science, unknown or unfamiliar to those exposed, invisible/unobservable, and having delayed effects. Dread risks are those judged as uncontrollable and involuntary, where the risks are inequitably distributed, increasing, and not easily reduced, potentially impacting large numbers of people around the world or affecting future generations, and with consequences that are fatal, dreaded, and catastrophic. A third factor, reflecting the number of people potentially exposed to the risk, has also been found in a few studies (Slovic et al., 1985).

Research has shown that experts typically judge levels of risk in terms of the probability of some event that is seen as undesirable combined with some assessment of its expected harm. Thus, experts often judge riskiness in terms of expected annual morbidity or mortality (Slovic et al., 1979). In contrast, while laypeople's judgments of risk are also sensitive to such assessments of the expected probability and magnitude of harm, research has also demonstrated that people's risk perceptions and attitudes are also strongly related to the hazard's characteristics as defined by the unknown and dread risk factors. Most important is the dread risk factor (Slovic, 1987). In general, people see hazards with higher scores on this factor as posing a greater risk. They are also more likely to judge the risk as unacceptable, to want to see the risks of the particular hazard reduced, and are more willing to support strict regulation to achieve a reduction in risk (Slovic et al., 1985). Sparks and Shepherd (1994a) found that microbiological contamination was rated high on the dread factor but low on the unknown factor. However, specific examples of microbial contaminants (*Listeria* and *Salmonella*) were rated higher on the dread factor than was a more generic description of the hazard of "bacterial contamination." This may explain why, in their advertisements, some marketers choose to include specific examples of the kinds of germs their products are designed to kill.

For risk communicators, this line of research offers at least two important lessons. The first is that conflicts between experts and laypeople regarding risks and what to do about them can be rooted in their differing definitions of what constitutes an acceptable or unacceptable risk (Slovic, 1987). The results also help to explain the apparent "irrationality" of laypeople's comparative judgments of risks (Wandersman and Hallman, 1993). For example, the influence of the "unknown" and "dread" characteristics of hazards in people's judgments about risks helps to explain why new microbial hazards with rather low probabilities but dreadful consequences like infection with Ebola, SARS, or avian influenza can generate fear, while more common but familiar microbial risks such as those posed by seasonal influenza or food-borne bacteria are often ignored. It also helps to explain why the identical risks can be judged as completely unacceptable if they are felt to be imposed by others, while they are seen as inconsequential if under one's own control. For example, laypeople are likely to see poor hygiene practices at a restaurant as completely unacceptable (and disgusting), while simultaneously engaging in variations of the same unhygienic practices in their home kitchens. The important lesson is that risk communicators must analyze the situational characteristics of the microbial hazards they are confronting and should design their messages accordingly (Miles et al., 1999).

CULTURE AND FOOD RISKS

While risk communication research involving other types of hazards can inform efforts to design communications about microbial risks, concerns about food risks also have unique characteristics that should be considered independently of perceptions of hazards in other domains (Fife-Shaw and Rowe, 1996; Frewer et al., 1997; Sparks and Shepherd, 1994b). For example, food is laden with religious, symbolic, and cultural meanings that set it apart from other concerns. Indeed, anthropologists have found that feasting, fasting, and the ritual preparation and consumption of certain foods, and taboos or restrictions regarding the touching or eating of other foods all play central roles in religious and cultural practices and identities (see, for example, Bynum, 1985; Douglas, 1966, 1972; Fiddes, 1994; Lévi-Strauss, 1966, 1970).

Food and food preparation can also be used as an expression of personal creativity, and the consumption of particular foods and rejection of others can also be seen as part of establishing or communicating one's personal identity, role in society, or political or ideological viewpoints (Sadalla and Burroughs, 1981). Therefore, risk communicators need to be aware that critiques of food preparation or consumption practices (what people do) may be perceived as criticisms of who people *are*.

The provision or sharing of food is also symbolically, psychologically, and emotionally linked with love, nurturing, and intimacy, and is considered critical to creating and maintaining bonds between people (Douglas, 1966; Miller et al., 1998). Because of this, providing contaminated or unhealthy foods to someone else may be considered particularly offensive. As such, risk communicators need to keep in mind that there are stigmas attached to being accused of unsanitary practices, giving or purveying spoiled, unclean, or unsafe foods to others, or for making others ill (Harris, 1985). As a result, some people may deny or refuse to take responsibility for errors of omission or commission that result in these outcomes. So, risk communicators must be tactful in how they present information dealing with these issues. At the same time, they may also want to use the aversion people have toward being stigmatized as a motivation to change behaviors.

Risk communicators also need to be aware that certain foods carry special symbolic importance (Douglas, 1966). For example, the contamination of foods such as milk, honey, or apples that are culturally associated with purity and wholesomeness may be seen as especially abhorrent. Therefore, the perceived risks associated with their adulteration may be enhanced because of their symbolic value (Haddix, 1990). Similarly, the corruption of foods associated with feeding infants or children may be seen as particularly risky both because of the perceived vulnerability of the potential victims and the symbolic violation of the means by which children are nurtured (Harrison, 2001).

AFFECTIVE RESPONSES TO FOOD RISKS

Communicators about microbial risks need to understand that perceptions of and reactions to the microbial contamination of food can evoke core affective responses. Slovic et al. (2004) argue that in addition to using an "analytical system" to judge risks, people also rely on an intuitive "experiential system" that is rapid, automatic, and not necessarily part of conscious awareness, "relying on images and associations, linked by experience to emotion and affect (a feeling that something is good or bad)." They further suggest that this experiential system has evolutionary roots and that representing a risk as a feeling remains "the most natural and most common way to respond to risk."

In contrast to proponents of formal risk analysis who tend to view affective responses to risk as irrational, Slovic et al. (2004) argue that the analytical and the experiential systems operate in parallel and that each is guided by the other. They conclude that analytical reasoning cannot be effective unless it is guided by emotion and affect and that "rational decision making requires proper integration of both modes of thought." Risk communicators must therefore have a good understanding of likely affective responses to food risks, especially triggers of disgust.

Because they are ingested, food risks represent hazards that are perceived as being both physically and symbolically incorporated into one's corporal being. Since "you are what you eat," consuming impure foods results in an unclean body (Angyal, 1941). As a result, potential food hazards are often seen as objects of disgust, characterized by revulsion at the prospect of oral incorporation of an offensive and contaminating object (Angyal, 1941; Rozin and Fallon, 1987).

Disgust was counted by Darwin (1872) as one of the six basic human emotions and is considered to be a powerful, culturally universal human reaction offering evolutionary advantages in the avoidance of illness (Rozin et al., 2000) or infectious disease (Curtis et al., 2004; Curtis and Biran, 2001). However, it is also necessary to understand that there are several potential motivations for rejecting foods (Fallon and Rozin, 1983). Rozin and Fallon (1987) suggest that some things may not be eaten because they are of minimal nutritional value and in general not considered to be "food" within a particular culture. Some foods may be disdained on the basis of sensory-affective reactions to bad tastes, odors, or textures; they are considered edible within a culture but are not eaten by an individual as the result of personal preferences. As has already been suggested, certain foods may be rejected based on the anticipation of harm following their ingestion. However, some rejections based on anticipated danger may be situational; for example, a food normally eaten might be rejected simply because it has passed its "sell by" date. Some foods, such as poison mushrooms, may be considered universally dangerous, while food allergies or intolerances might make some foods hazardous only to specific individuals.

Finally, some foods may be spurned because they are judged disgusting based on knowledge of the origin or nature of the food or on the basis of who (or what) may have touched it. Such items have offensive properties and are presumed to taste bad (though most people have never tasted many of the things they find disgusting). They also have the power to contaminate other things. Both Angyal (1941) and Rozin and Fallon (1987) note that most of the things found to be disgusting are associated with animals or animal products, with feces being the prototypical example. However, Curtis et al. (2004) present evidence that the association of an object or substance with a potential disease threat makes it more disgusting, suggesting that the emotion of disgust may be an evolved response to a variety of objects in the environment that represent threats of infectious disease. This is consistent with Rozin and Fallon (1987) who argue that because spoiled or decayed items may be disease vectors, these particular substances may form the core of disgusts. Martins and Pliner (2006) suggest that associations with "animalness" and aversive textural properties such as mushiness or sliminess (potentially indicating spoilage) underlie judgments of disgust. There is also evidence that consumers use such "disgust" indicators to make judgments about food safety. For example, in a national survey of 1,025 adults in Ireland (McCarthy et al., in press), more than half (54%) reported that "You can always tell that a food is unsafe to eat by the appearance and smell of the food."

Rozin and Fallon (1987) also argue that an important aspect of disgust is the phenomenon of psychological contamination, whereby past physical contact between a contaminant and an otherwise acceptable food item causes the food to be rejected as disgusting. They suggest that a critical issue in understanding contamination is the importance of physical contact between the adulterant and that which has been adulterated and people's belief that after that contact a physical trace of the contaminant may persist. As such, even when something disgusting (like an insect) is removed from a food item, people insist that a physical trace of the insect may remain, rendering the food unacceptable even if the trace is undetectable (Fallon et al., 1984; Rozin and Fallon, 1980). As Rozin and Fallon (1987) point out, this phenomenon follows the law of "sympathetic magic" originally described by James Frazer (1890) and Marcel Mauss (1902) to explain a variety of magical beliefs and practices in traditional cultures. A fundamental part of the law, which Frazer refers to as "contagion," encapsulates the belief that objects that come into contact with others continue to exert an influence on each other even long after they have been separated. Thus, foods that come into contact with something (or someone) considered undesirable may be stigmatized as contaminated forever. For example, some people refuse to use plates, dishes, glasses, or silverware that were once used by people with infectious diseases such as HIV, even if those objects have been washed and sterilized (Rozin et al., 1992). Alternatively, contagion can work in a positive direction. For example, foods can take on desirable qualities simply because they were made by a loved one (Miller et al., 1998). Ironically, this may be the case even if the food was made under unsanitary conditions and may, in fact, be unknowingly contaminated.

A second important aspect of the law as described by Mauss (1902) is that the tiniest part of an object embodies all of the attributes of the whole object. In the practice of magic, personal residues such as hair, fingernail clippings, and so on, are thought to retain properties of their original owners and so are useful for the purposes of sorcery. In the practice of food preparation, similar residues may be seen as disgusting both because they can be vectors for harmful disease *and* because they might convey some unsavory attributes of the person to whom they belonged.

Another of the fundamental principles of sympathetic magic as described by Mauss (1902) is that of similarity. According to this principle, an image of, or resemblance to, an object captures a fundamental similarity or identity of that object. As a result, actions visited on the image are transmitted to the original object (or person). This principle helps to explain why, in some cases, disgust can be evoked by objects that only look like, or are associated with, a disgusting item. For example, in an experiment designed to test this idea,

Rozin et al. (1986) found that even when they knew what they were ingesting, people much preferred consuming a piece of chocolate fudge shaped like a muffin than eating a piece of the same fudge shaped in the form of dog feces. Similarly, Rozin et al. (1984) found that stirring a favorite soup with a brand-new comb or fly swatter rendered the soup unacceptable for many respondents.

While it is unlikely that such imaginative confections would find commercial success, or that food preparers would be tempted to stir soup with a comb or a fly swatter, the point is that even if they are sterilized, objects typically associated with contamination can lead to perceptions of adulteration if they come into contact with "clean" foods. For example, the same type of plastic barrels commonly used as trash receptacles are also sometimes used for bulk food or ingredient storage; for some, the idea of storing food in "trash cans" might be considered disgusting.

PUBLIC PERCEPTIONS ABOUT THE RISKS OF MICROBIOLOGICAL HAZARDS IN FOOD

It is important for risk communicators to understand that while consumers may find microbial contamination of food to be disgusting, they also tend to underestimate the likelihood of it affecting them. While food safety experts judge microbiological hazards to be the main risk to health from food, consumers rank the risk of microbiological hazards as considerably lower than those posed by pesticides and food additives (Brewer et al., 1994).

In a national survey, Lin et al. (2005) found that the majority of U.S. consumers said they had heard of *Salmonella* (94%) and *Escherichia coli* (90%) as a problem in food. Yet, only about one-third were aware of *Listeria* (32%) and only 7% were aware of *Campylobacter*. Sixty-two percent thought it was very common or somewhat common for people in the United States to become sick because of the way food was handled or prepared in their homes. However, only one in three consumers thought they or someone in their household had been sick from eating spoiled or unsafe food.

In fact, Altekruse and Swerdlow (1996) point out that the epidemiology of food-borne diseases in the United States has changed in the past few decades as the result of the emergence of new pathogens, changes in both the types of food that people eat, and the sources of their food, and an increase in the number of people who are particularly susceptible to food-borne illnesses. They argue that changes in the global economy have made possible the rapid shipment of perishable produce and other foods, increasing the likelihood of exposure to food-borne pathogens from other parts of the world. They also argue that extensions in life expectancy both for the average citizen and for the chronically ill and the immunocompromised have increased the proportion of the population susceptible to severe illness after infection with a foodborne pathogen. All of these changes make awareness of microbial risk and good hygiene practices more important, while evidence suggests that public awareness of safe food preparation practices may be declining.

WHAT DO PEOPLE DO?

Developing appropriate messages regarding microbial risks requires an understanding of people's behaviors. That consumers make frequent common mistakes in their domestic food-handling and food-consumption practices is not in doubt. Home food-preparation practices have been investigated using a variety of techniques, including telephone (e.g., Altekruse et al., 1996; Woodburn and Raab, 1997) and mail surveys (e.g., Williamson et al., 1992), focus groups and face-to-face interviews (e.g., Spriegel, 1991), and observational studies of participants as they prepared meals (e.g., Daniels, 1998; Jay et al., 1999). Some studies have also taken microbiological samples from consumers' kitchens (Finch et al., 1978; Scott et al., 1982; Sharp and Walker, 2003; Speirs et al., 1995).

As Sharp and Walker (2003) note, "All of these studies indicate that unhygienic practices occur frequently and identify sites where high levels of microbial contamination are likely." Indeed, in an observational study of home-based food-handling practices, Daniels (1998) found that 96% of the participants made at least one critical food safety mistake that could result in a food-borne illness. Similarly, an observational study of 108 consumers asked to prepare one of four standard recipes (Worsfold and Griffith, 1997a; Griffith et al., 1998) found that 95% failed to implement basic hygiene practices due to lack of knowledge or failure to implement known food safety procedures, making common mistakes in the transport, refrigerated storage, handling, preparation, and cooking of foods. They observed that 45% of those in their sample transported vulnerable raw ingredients at temperatures that were too warm for too long and 58% then stored these ingredients in refrigerators that were too warm. Two-thirds failed to wash their hands before beginning food preparation, 58% did not wash their hands after touching raw chicken or beef, and 76% did not wash their hands after cracking eggs. One-third washed raw poultry in the sink, 41% failed to wash vegetables, and 60% used the same cutting board to cut both. About 15% neglected to cook the recipe to the recommended internal temperature, more than a third (35%) let the cooked food remain at ambient temperature for more than 90 minutes, 17% stored the remaining food in a refrigerator with a temperature above 5°C, and 11% failed to reheat the leftovers to a temperature of at least 74°C.

A national survey in the United Kingdom (Foodlink, 2005) found that about three-quarters (76%) could correctly identify the reason why raw meat should be stored on the bottom shelf of the fridge (i.e., so that it will not drip onto other foods). However, only about half of the respondents (48%) knew the correct temperature for the inside of a refrigerator (between 0 and 5°C). Not surprisingly, in an earlier observational study of 108 food preparers in homes in the United Kingdom, 58% of those in the study stored chilled ingredients in a refrigerator that operated above 5°C (Worsfold and Griffith, 1997a). Similarly, Johnson et al. (1998) studied food safety knowledge and practices among 809 elderly people (aged 65+) living at home in urban Nottingham, United Kingdom. They found that 70% of the refrigerators in the homes of those studied were too warm for the safe storage of food ($\geq 6°C$).

Of particular importance are the hazards stemming from cross-contamination. In an observational study of the preparation of a popular chicken casserole recipe, Meredith et al. (2001) also found extensive cross-contamination in a model kitchen. However, the most frequently contaminated objects and sites were those associated with the cleaning process and the maintenance of personal hygiene, including dishcloths, towels, and the sink area. This is consistent with a number of studies that have shown that dish cloths and other fabrics may become both habitats and vectors for microbial contaminants (Scott and Bloomfield, 1990, 1993; Sharp and Parkinson, 1999) and that consumers often wipe their soiled hands on towels and aprons, and regularly use the same cloth for washing dishes and wiping surfaces (Ackerley, 1994; Redmond et al., 2004; Worsfold and Griffith, 1997a, 1997b).

To examine the microbiological consequences of unhygienic practices reported in previous observational studies, Haysom and Sharp (2004) prepared a simple meal using raw chicken and lettuce in a manner that incorporated examples of common food safety mistakes. They then traced the spread of contamination caused by preparing this meal. After meal preparation and cleaning, salmonellae were found to be present on at least one surface in the kitchen in 97% of the 30 meal preparations. Of the prepared meals, 90% were positive for salmonellae.

Sharp and Walker (2003) note that the risk of cross-contamination may be exacerbated in particular in communal kitchens owing to the number of individuals using the kitchens, confined space, lack of feelings of responsibility, and differing standards of knowledge and hygiene. They also point out that food safety in such communal arrangements may be compromised by the actions of just one of the individuals using the kitchen. Of course, the same is true of commercial food preparation kitchens.

In fact, a telephone survey of more than 8,000 food service workers in the United States examining self-reported food-preparation practices in restaurants suggests that risky behaviors that can result in food-borne illness are common (Green et al., 2005). Sixty percent of the respondents reported that they did not always wear gloves while touching ready-to-eat (RTE) food. Nearly one-quarter (23%) did not always wash their hands, and one-third (33%) did not always change their gloves between handling raw meat and RTE food. More than half (53%) did not use a thermometer to check food temperatures, and a minority (5%) report that they had worked while sick with vomiting or diarrhea.

Hand Washing
Because of the importance of hand hygiene in preventing the spread of disease, extensive efforts have been made to encourage proper hand washing. As part of these campaigns, widespread surveillance efforts have been carried out in the United States and the United Kingdom. The results show that despite modest success of public health campaigns designed to improve hand hygiene, the current rates and practices of hand washing are inadequate, women wash their hands more often than men do, but neither wash their hands as often as they think (or say) they do.

In a national survey by the British Food and Drink Federation and the Institute of Environmental Health Officers (1995), 93% of respondents emphasized the importance of hand washing in food-preparation situations. In an update of this study (Foodlink, 2005), 8 of 10 (81%) Britons said they always wash their hands before preparing food, and almost 9 of 10 (88%) say they always wash their hands after handling raw meat, poultry, or fish. However, while people *say* they wash their hands, in practice, however, Worsfold and Griffith (1997b) found that 64% of those observed preparing food did not wash their hands when starting food preparation and 47% did not wash their hands at all.

In addition, some consumers' versions of what constitutes hand washing fall short of acceptable practice. The British Foodlink (2005) study found that while people say they are washing their hands more often, 29% reported that they do not always use soap and 28% do not always dry their hands properly after washing. In part, this may because when asked whether damp or dry hands were most likely to spread germs, only 79% correctly said damp; 15% said dry and 6% said they did not know. Many also do not have a clear sense of how long germs remain viable on one's hands; less than one-quarter (23%) correctly identified that germs could survive for up to three hours on your hands.

When asked why people did not wash their hands before preparing food, more than one in five (22%) said they were too lazy; 21% said they did not care or think about it; more than one in 10 (12%) cited forgetfulness; 8% stated lack of time, and 3% cited "hands looking clean" as a reason for avoiding soap and water. The same survey also found that nearly half (47%) of Brits report that they do not always wash their hands before eating lunch at work.

After interviewing more than a thousand workers and managers in small independent catering businesses in the United Kingdom, the Food Standards Agency (2002) concluded that although there was a general understanding among all workers that they should wash their hands (64%), nearly four in ten (39%) neglected to wash their hands after using the toilet and half (53%) did not appear to wash their hands before preparing food. The study also found that only 32% of the workers believed good food hygiene practices were important to their business compared with 64% who saw good food as the key to keeping their customers. Moreover, while almost all of the catering managers interviewed agreed that food poisoning can be life threatening, only 42% put food hygiene on their priority list for the success of their business. Not surprisingly, in the Food Standards Agency's 2002 annual "Consumer Attitudes Survey," more than half of all consumers (51%) expressed concern about the standards of hygiene in catering businesses (Food Standards Agency, 2003).

In the United States, rates of hand washing are scarcely better. In a study sponsored by the American Society for Microbiology (ASM) in 2005, 91% of 1,013 American adults interviewed by telephone reported that they always wash their hands after using public restrooms. However, only 83% actually did so, according to a separate observational study of 6,336 individuals at six public attractions in four major cities: Atlanta (Turner Field), Chicago (Museum of Science and Industry, Shedd Aquarium), New York City (Grand Central Station, Penn Station), and San Francisco (Ferry Terminal Farmers Market). Ninety percent of the women observed washed their hands, compared with 75% of men.

The ASM telephone survey also found that most Americans say they always wash their hands after such activities as using the bathroom at home (83%), before handling or eating food (77%), and after changing a diaper (73%). However, only a minority said that they always washed their hands after petting a dog or cat (42%), after handling money (21%), or after coughing or sneezing (32%). There were clear gender differences in this latter behavior. While 39% of women say they always wash their hands after coughing or sneezing, less than one-quarter (24%) of men reported the same practice.

An earlier (2003) observational study sponsored by the ASM found that many people passing through major U.S. airports do not wash their hands after using the public toilets. More than 30% of the 7,541 people observed using airport restrooms in New York, Chicago, San Francisco, Dallas, Miami, and Toronto did not stop to wash their hands. Again gender differences were apparent. Overall, 83% of the women were observed washing their hands, while only 74% of their male counterparts stopped to wash their hands. However, hand washing was nearly universal at the Toronto airport, a city that had recently witnessed a major outbreak of SARS; 97% of women and 95% of men were observed washing their hands, likely heeding advice that one way to avoid SARS is scrupulous hand washing.

Even among health professionals (who should be aware of its importance) hand washing frequency and technique often fall short. In an observational study of 120 nurses providing patient care, O'Boyle et al. (2001) found that the nurses adhered to hand washing guidelines only 70% of the time. The correlation between self-reported and observed adherence to hand-washing recommendations was low ($r = 0.21$). Similarly, in their study of 60 nursing students Snow et al. (2006) found that while the students had strongly positive attitudes toward washing hands and wearing gloves, they had an overall low rate of hand hygiene. Male students practiced hand hygiene 30% less often than their female counterparts.

WHO IS MOST LIKELY TO ENGAGE IN UNSAFE PRACTICES?

As with studies examining hand-washing behaviors, researchers have found demographic differences in risky food-handling and -consumption behaviors. Understanding these differences is important to properly target specific audiences with tailored risk communication messages.

In a study designed to identify high-risk populations, Altekruse et al. (1999) used Behavioral Risk Factor Surveillance System (BRFSS) telephone interviews of 19,356 adults in eight states (1995: Colorado, Florida, Missouri, New York, and Tennessee; 1996: Indiana, New Jersey, and South Dakota) that included questions related to food-handling and food-consumption practices. Risky food-handling and food-consumption practices were common among the survey respondents. Overall, 19% of respondents did not adequately wash hands or cutting boards after contact with raw meat or chicken. During the previous year, one in five (20%) ate undercooked hamburgers, half (50%) ate undercooked eggs, 8% ate raw oysters, and 1% drank raw unpasteurized milk. Men and young adults were more likely to report risky food-handling and food-consumption behaviors. They also found that consistent with an earlier nationwide survey conducted by the Food and Drug Administration (Klontz et al., 1995) the prevalence of most risky behaviors increased with increasing socioeconomic status.

Klotz et al. (1995) conducted telephone interviews with 1,620 American adults. They found the safest food-preparation and -consumption practices were reported by women, those older than 39 years, and those with a high school education or less. However, they found that many Americans reported consuming raw animal-based foods. More than half (53%) reported consuming products with raw eggs; nearly one-quarter (23%) reported eating undercooked hamburgers; 17% ate raw clams or oysters; and 8% had consumed raw sushi or ceviche. Twenty-five percent of the respondents said that after cutting raw meat or chicken, they used the cutting board again without first cleaning it.

Zhang et al. (1999) reported the results of a survey conducted between 1995 and 1996 in Kansas. They showed that more than half (55.6%) of their respondents reported eating raw or undercooked eggs. This was most common among those with teenagers and with higher levels of education. They also found that more than one-quarter (26.5%) of their respondents consumed home-canned vegetables, with those living in rural areas and those with teenagers most likely to do so. A minority (8.7%) reported eating undercooked hamburger, but those who were overweight were more likely to do so, while those with a child under the age of 5 were less likely to do so. Only 1.8% of respondents consumed raw milk, with higher prevalence among lower-income respondents.

In their national telephone survey of Americans, Altekruse et al. (1996) found that unsafe food-handling practices were reported more frequently by men than by women, by people aged 18 to 29 years compared with people aged 30 or older, and by people who prepared food infrequently as compared with frequent food preparers. Those with lower rates of self-reported safe practices had a level of food safety knowledge similar to that of the sample overall.

In their study of self-reported practices of food handlers in restaurants, Green et al. (2005) found several factors were associated with safer food-preparation practices. For example, workers whose jobs included responsibility for preparing food reported washing their hands and wearing gloves when handling foods more often than workers not responsible for food preparation did. Workers involved in cooking reported changing their gloves more often than did workers who did not cook. Older workers and managers reported washing their hands more often than younger workers and line employees did. They also found that workers in chain restaurants were more likely to report using thermometers to check food temperatures than were workers in independently owned restaurants.

WHY DO PEOPLE FAIL TO PRACTICE GOOD FOOD-PREPARATION HYGIENE?

Altekruse and Swerdlow (1996) suggest that there is evidence of a possible decline in public awareness of safe food-preparation practices. For example,

Williamson et al. (1992) found that many Americans lack a clear understanding of the types of organisms that cause food-borne illnesses and have inaccurate ideas about what foods were particularly at risk. They also underestimate the importance of cooking food to proper temperatures and the need to avoid cross-contamination.

However, lack of knowledge may only be part of the answer. A larger problem seems to be getting people to use the knowledge they already have. In a review of 87 consumer food safety studies using a variety of methods, Redmond and Griffith (2003) conclude, not surprisingly, that studies that rely on self-reported survey data typically provide a more optimistic representation of consumer food safety behaviors than those using data collected by direct observation. They note that "although consumers have demonstrated knowledge, positive attitudes and intentions to implement safe practices, substantially larger proportions of consumers have been observed to implement frequent malpractices." In other words, people often know what they are supposed to do regarding food safety, acknowledge the importance of such practices, but then fail to implement them.

As an illustration, in their national telephone survey of U.S. residents, Altekruse et al. (1996) reported that their respondents were familiar with food-handling principles but failed to employ them. For example, while 86% knew that hand washing reduced the risk of food poisoning, only two-thirds reported washing their hands after handling raw meat or poultry. Eight in ten knew that serving cooked steak on the plate that had previously held raw meat increased the risk of food poisoning, but again, only two-thirds said that they cleaned a cutting board after contact with raw meat or poultry.

Similarly, in their study of knowledge, intentions, and behaviors of 40 United Kingdom consumers who prepared food in the home, Clayton et al. (2003) found that, despite the fact that *all* of the participants answered questions about hand washing correctly and 85% said that they were very likely to wash their hands after handling raw ingredients, *none* of the participants washed their hands at all of the appropriate times when observed preparing a meal. Similarly, 100% of the participants were aware of the need to either wash or use different utensils between contact with raw ingredients and cooked products, yet fewer than half (48%) did so at all times. Moreover, nearly three-quarters (73%) knew both how to most effectively clean surfaces and equipment and when it was appropriate to clean, yet only 40% adequately carried out cleaning activities at all times.

Clayton et al. (2003) note the available research clearly suggests that while knowledge may be an important antecedent to effective food safety practices, knowledge by itself is not a good guarantor that such practices will be implemented. For example, in a study of professional food handlers, Clayton et al.

(2002) found that, despite the fact that 95% of their respondents had received food hygiene training, 63% admitted to sometimes not carrying out appropriate food safety behaviors. As such, food safety education courses, while potentially increasing knowledge, may not necessarily lead to changes in practice among consumers or professionals (Howes et al., 1996; Powell et al., 1997). In fact, in a review of studies evaluating food hygiene education programs for professional food handlers and managers, Rennie (1994) concluded that there is a lack of evidence of improved food hygiene standards resulting directly from training programs alone. She argued that despite a long history of such training programs, many evaluations of food safety certification courses in both the United States and the United Kingdom indicate the "failure of formal courses to generate improvements in food handling practices."

As Worsfold and Griffith (1997a) suggest, "Ignorance may not be the major problem. People may fail to apply already well-known principles. The real challenge for hygiene education is to persuade people to translate what they know into practice. The problem of changing people's behaviour is complex. Unhygienic practices, often deeply ingrained habits, are not easily displaced, even by the most imaginative teaching programmes."

So, people tend to overestimate their knowledge of, and control over, the risks of microbial contamination, they exaggerate their fidelity in following food safety hygiene practices, and they often consciously violate what they know to be best practices. The question is why?

There are several probable situational, attitudinal, and psychological factors involved. For example, despite knowing best practices, commercial food handlers point to a lack of time, space, staff, utensils, supplies, equipment, and management support as significant barriers to carrying out necessary food safety procedures (Clayton et al., 2002). As such, Rennie (1994) points out that, while food safety hygiene courses may increase knowledge about best practices, participants may lack the resources or authority to make necessary changes, especially if they involve purchasing equipment or making structural improvements in physical facilities.

In a national survey of Irish consumers, McCarthy et al. (2007) compared the salient beliefs of consumers toward food safety with observations of their food preparation behaviors. They conclude that the clear majority know what they should be doing when preparing food in the kitchen. However, the results suggest that, in many cases, people *also* regard less-than-ideal food-handling practices as safe. For example, while 71% correctly identified that meat should be safely defrosted in the refrigerator, 61% also felt that defrosting meat on the kitchen counter was safe. Thus, most people can readily identify best practices, but they may not believe that alternative practices are significantly less safe. As McCarthy et al. suggest, "It is most likely that many have considerable experience of defrosting meat on a kitchen counter and believe that they have not experienced any illness as a result; thus, breaking a habit could prove difficult unless there is ample evidence to support this change in behavior."

Similarly, Clayton et al. (2003) suggest that one reason people do not practice the food safety precautions they know about is that they do not believe that failing to carry them out is likely to pose much risk to themselves or to others. In part, this may be because they do not connect faulty food safety practices with their consequences, arguing that "I have not given anyone food poisoning yet, so I cannot be doing that much wrong."

Optimistic Bias Effects

The idea that "it hasn't happened yet, so it won't in the future" is consistent with the idea that food-borne illnesses may be prone to optimistic bias effects whereby individuals believe that they are less at risk from a hazard than other people (Weinstein, 1984, 1987). Weinstein (1987) found that an optimistic bias is often introduced when people estimate their future vulnerability to a hazard based on extrapolations from their own past experiences. Thus, the hazards most likely to elicit unrealistic optimism are those associated with the belief (often incorrect) that if the problem has not yet occurred, it is unlikely to do so in the future. Optimistic biases also increase with the perceived preventability of a hazard and decrease with perceived frequency and personal experience.

However, in the case of food-borne illness, this reliance on personal experience may be particularly problematic since people often do not recognize the symptoms of food poisoning (Leman, 2001) and underestimate the incidence of food-borne illness (Hayes et al., 1995), so they do not connect their poor food-preparation hygiene practices with their consequences. Woodburn and Raab (1997) found that the majority of their respondents could not identify groups of people particularly at risk for food poisoning and had difficulty identifying the symptoms of food-borne illness. Respondents typically focused on symptoms of gastrointestinal upset, ignoring fever and other signs of illness. They also did not recognize that many food-borne illnesses have an incubation period longer than 24 hours, believing that such illness typically occurs within a day after the contaminated food was eaten. This may explain both people's inability to connect their actions with the consequences and the low incidence of reporting of food-borne illnesses. However, Parry et al. (2004) found in a case-control study that people who had recently experienced *Salmonella* food poisoning rated their personal risk from food poisoning as higher than controls did, suggesting that personal experience of food poisoning can reduce optimistic bias.

That food-borne illnesses are underreported is in little doubt. Palmer et al. (1996) compared the self-reported rates of gastroenteritis and suspected food poisoning with the actual rates of patient visits for these problems within the same medical practice population. They found that while 0.3% of the practice population visited their doctor for problems with gastroenteritis and 0.06% saw their doctor as the result of suspected food poisoning each month, 7% reported having had symptoms of gastroenteritis and 0.7% suspected that they had suffered from food poisoning. Thus, they estimated that for every patient who had seen a doctor for such an illness, there were 26 others who had not. This suggests that even many of those who *suspect* they are suffering from food poisoning do not see their doctors and their cases are unlikely to come to the attention of health authorities. Thus, official tallies of food-borne illnesses are likely to significantly underestimate the true rates of infection within the population.

Fein et al. (1995) used two FDA surveys conducted in 1988 and 1993 to characterize consumer perceptions of food-borne illness. They found that younger people (18 to 39 years of age) were more likely than those in other age groups to believe they had experienced a food-borne illness. Better educated people (with at least some college education) were more likely to believe they had experienced food-borne illness than were people with less education. Those who thought they had experienced a food-borne illness reported a greater awareness of food-borne microbes and greater concerns about food safety issues. However, they were more likely to eat raw protein foods from animals (such as raw egg products) and less likely to practice safe food handling than were those who did not perceive that they had experienced such an illness.

Weinstein (1984) also suggests that risks people believe are preventable by personal action, or are under their own control, are likely to evoke unrealistic optimism. Not surprisingly, unrealistic optimism and illusions of control concerning food preparation in the home have been found to be prevalent (Miles and Scaife, 2003; Redmond and Griffith, 2004). Data from the United Kingdom (Frewer et al., 1994) suggest that consumers feel they have relatively high levels of control over the prevention of food-borne illness in the home and that the greatest risk comes from food prepared by others. They also conclude that many consumers feel they have a knowledge of hazards that is greater than the knowledge other people have. This combination of high perceived knowledge and control may mislead people into believing that they know enough about the potential hazards associated with foods prepared in the home to deal with them effectively.

Miles and Frewer (2001) conducted in-depth interviews with 26 participants, finding that concerns about health and the risk of death were the main issues their respondents associated with *Salmonella*. Those interviewed identified eggs and chicken as particularly risky foods and had some awareness of ways to prevent food poisoning by *Salmonella*, including proper cooking, storage, and defrosting precautions. They reported that they felt safe when cooking for themselves because they followed these preventive measures, but they were concerned about eating out.

Similarly, in a study of the beliefs, attitudes, intentions, and behaviors of 40 United Kingdom consumers who prepared food in the home, Clayton et al. (2003) found that consumers generally agreed with the statement "food poisoning is a disease which can result in very serious consequences." Yet, three-quarters of these same consumers also said that it is unlikely that they would get food poisoning in their home and discounted the likelihood that the food that they prepared in their homes would cause others to get food poisoning.

In fact, consumers typically underestimate the percentage of food poisoning cases that originate in faulty home food-preparation practices, believing that the major responsibility lies with food manufacturers or restaurants (Worsfold and Griffith, 1997a). For example, Williamson et al. (1992) reported that only 16% of consumers thought that food safety problems were common at home, while one-third thought that food safety problems were most likely to occur at food-manufacturing facilities and one-third thought that such problems were most common in restaurants. Altekruse et al. (1996) reported similar results. In their study, only 17% identified food poisoning as commonly originating in the home, 17% blamed supermarkets, and 65% held restaurants most responsible.

Unrealistic optimism about the risks of causing food poisoning is not restricted to food preparers in the home. Clayton et al. (2002) studied the beliefs and self-reported practices of commercial food handlers selected from businesses preparing high-risk foods in Wales. Ninety-five percent of the respondents had received training in food hygiene. So, while the food handlers were aware of proper food safety practices, 63% admitted that they did not always perform them. Yet, consistent with findings from earlier research (Coleman and Griffith, 1998; Mortlock et al., 1999), the majority believed that there was a low risk of someone contracting food poisoning from their business. In fact, they judged the risks of food-borne illnesses to be greater at businesses similar to their own and the risk at their own homes to be greater still.

Miles et al. (1999) suggest that public perceptions of food-borne illnesses may also be prone to optimistic bias effects because most risk information communicated to the public is about risks to people in general. Individuals must then use this information to judge their own specific risk status. In doing so, people have a tendency to underestimate their own risks relative to others, creating a gap between their perceived risks and their actual risk status. As a result, people may ignore risk communications, assuming that the messages are aimed at other more vulnerable individuals.

GETTING YOUR MESSAGE ACROSS

The Importance of Message Repetition

Keep in mind that marketing experts have learned that there are limits to what an audience can or is willing to absorb after a single exposure to a message. Therefore, repetition of key messages can be important to ensure that people pay attention to and retain important ideas. For example, refresher courses and ongoing reinforcement of food hygiene training within the workplace can improve overall knowledge among food preparers (Rennie, 1994).

Unfortunately, those key messages are unlikely to be seen on television news programs, where most people say they turn to learn about science and current events (National Science Board, 2002). In the United States, Nucci et al. (2006) found only 185 stories related to food safety issues reported on the three major network news programs over a 6-year period between 2000 and 2005, an average of only 10 stories per network per year. Moreover, much of the consistent news coverage of food safety issues during this time focused on mad cow disease. They concluded that the average viewer would learn very little about food safety issues from the nightly news. The stories provided little or no information about proper food-preparation techniques, and there was very limited dissemination of information about food recalls and advisory notices. When stories covering recalls did air, there was no information, such as lot numbers or the extent of the recall, that would help consumers identify potentially affected products or instructions about how to return suspect items.

Choosing Appropriate Channels for Your Messages

Embedding key food safety messages within an appropriate context and proximal to the situations where they can be acted on is also important. For example, magnets listing proper cooking temperatures for meats and refrigerator/freezer thermometers clearly indicating when temperatures are out of the safe range are simple devices that can provide important, actionable information at or near the site where it is most needed.

Food labels can serve a comparable role, though they often provide information to consumers but fail to explain why or how to use it. For example, Brandt et al. (2003) report that nearly one-third of the food products sold in the United States bear labels that include statements about the need for refrigeration. But the words "to maintain safety" are typically not present, even though FDA guidance indicates the importance of including them. As a result, it is not clear whether refrigeration is required to maintain food quality or food safety.

Similarly, about 7% of products sold in the United States bear a warning not to purchase a package with a broken or missing seal, or a statement disclosing that a safety button pops up when the original seal is broken. Example statements include "sealed for your safety," "do not purchase unless cap and neck band are intact," "do not use if seal under cap is broken," "do not purchase if the safety button is up" (Brandt et al., 2003). Yet, while providing important information and advice to consumers, none of the statements indicate *why* they should not use a product with a broken seal.

Likewise, while there is no uniform or universally accepted food product dating system in the United States, many food products voluntarily bear date stamps or carry expiration, "use by" or "sell by" dates without indicating their purpose. Though these are meant to explicitly address food quality and not food safety issues, many consumers believe that the dates indicate whether the food is safe to eat and use them as cues to discard packages (Brandt et al., 2003). Moreover, a 2002 Food Marketing Institute survey found that while nine of ten consumers reported being aware of these dates on food packages, and 88% said that they frequently or always look for them, most could not distinguish the differences between "sell by" and "use by" dates. There are also gender, age, educational, regional, national, and cultural differences in the extent to which people use these labels (Harcar and Karakaya, 2005). And, as Worsfold and Griffith (1997a) found in their study of food safety practices among the elderly, many experience difficulty reading food labels because of the small size of the text or the general need for reading glasses. As such, "use by" and "sell by" dates may not guide purchasing or storage decisions for most of the elderly.

Similarly, in a study of consumer preferences for sources of food safety information, Redmond and Griffith (2005a, 2005b) found that among consumers in the United Kingdom the most preferred source of food safety information identified was food packaging. Their respondents also reported that exposure to this source of advice was widespread, and 82% indicated that they were likely to read food safety advice stated on food packaging. However, like all static information, the food safety instructions on packages are often missed or ignored. As a result, Yang et al. (2000) suggest that the food safety advice on packages of raw meat is likely to have a limited influence on consumer practices in the home. Yet, such labels should be considered as only one component necessary to inform consumers about proper food-handling and -preparation practices, and to motivate people to change risky behaviors.

For example, Griffith et al. (1994) examined TV cooking shows, cook books, and popular cooking magazines for the information they conveyed about food handling and other food safety practices. They found that only a few cookbooks and magazines contained any information about food safety, and those that did mention food safety practices did so only briefly and inadequately.

They note that TV programs could be particularly effective in delivering messages about food safety. Indeed, many such programs are now hosted by a new breed of "celebrity chefs" who may be seen as role models, and TV cooking programs can visually demonstrate and reinforce the required skills and recommended practices necessary to prevent microbial contamination and resulting food-borne illness. However, they found that while the TV cooking shows they reviewed illustrated good personal hygiene practices, such as washing hands, they largely ignored other essential food safety issues, such as the need for thorough cooking, storage of foods at proper temperatures, and the prevention of cross-contamination. Moreover, Reid et al. (1998) point out that on many cooking programs food ingredients are often already prepared, so that good practices associated with raw foods are not always shown.

Similarly, Worsfold (1995) notes that many consumers use cooking magazines and cookbooks as sources of information on food and food-related issues. She cites a survey by McKie and Wood (1991) that found most respondents owned several cookbooks and many claimed to read them for pleasure. A study of women who read women's magazines found that more than half professed an interest in information about cooking, and many claimed to use the featured recipes in the magazines often or occasionally (Moore et al., 1992). Therefore, Worsfold (1995) urges food safety educators to persuade authors and editors of magazines, newspapers, and books that include recipes of the importance of providing specific food safety information for their readers within the published recipes.

She argues that every recipe should first be checked for the inclusion of ingredients such as chicken, turkey, eggs, shellfish, and so on, that might pose a food safety hazard. Then the preparation process detailed within the recipe should be assessed to determine whether it will effectively eliminate the risks posed by those ingredients. Worsfold emphasizes that because cooking foods to appropriate temperatures will kill a large proportion of food-borne pathogens, vague instructions to "cook until done" should be replaced with specific easy-to-follow instructions about cooking times and temperatures. She also emphasizes that because most consumers do not use thermometers to monitor temperatures during cooking, accurate timing information is very important. Moreover, she suggests that it would be helpful to include in recipes visual or other indicators that foods have been cooked thoroughly, such as changes in the color or texture of foods, or the flow or color of juices. Because many people cook meals in advance, she also suggests that recipes should provide specific guidance about how the cooked product should be treated if it is to be eaten later, including how to cool it, the length of time it can be stored, and how to safely reheat it.

While explicit instructions can be useful, appropriate reminders to people to practice what they already know can also be effective cues to action. For example, Thomas et al. (2005) tried to increase the rate of hand washing by "advertising" its importance in decreasing the incidence of health care-associated infections. Using 11 in. by 17 in. posters with visual cues developed with the periodic input of hospital personnel, compliance was increased over 12 months from a baseline rate of 20 to 37%. Realize, however, that most people quickly habituate to novel stimuli in their environments. This helps to explain the ineffectiveness of static messages such as the all-too-familiar poster over the washroom sink commanding "All employees must wash their hands." At the very least, posters with new designs and variations of an important message should be rotated and hung at frequent intervals.

Risk communicators must also give consideration to current consumer preferences and food consumption practices and adjust their messages and strategies accordingly. For example, an apparently widespread belief is that consumption of organic produce is safer, in part, because it has fewer microbial pathogens (Williams and Hammitt, 2001). Moreover, consumer preferences for the taste of raw seafood, "runny" eggs, and rare hamburgers, the need to minimize waste and make use of "leftovers," and the improved flavors of foods served at room temperature are powerful motivators to engage in unsafe behaviors. They are so powerful as to be unlikely to be undone by mere warnings, especially when people deny their own vulnerability, as is the case with many food and nutrition issues (Reid et al., 1998).

Connecting Actions to Consequences

To combat unrealistic optimism and the idea that "it can't happen to me," risk communicators need to take advantage of opportunities to connect people's actions with their consequences and to increase people's perceived vulnerability. Of course, part of the problem with microbial risks is that they are usually invisible, so it makes connecting actions and consequences difficult. As a result, some training programs designed to teach people to reduce cross-contamination and to promote better hand hygiene try to make "germs" visible by using luminescent materials such as Glo Germ or GlitterBug that glow under ultraviolet light.

Risk communicators also need to take advantage of events or circumstances that may increase people's sense of vulnerability or that can naturally focus people's attention on a difficult problem or issue, creating a window of opportunity for learning. Such "teachable moments" can be connected to salient personal experiences. For example, Parry et al. (2004) found in a case-control study that people who had recently experienced *Salmonella* food poisoning rated their personal risk from food-borne illness as higher than that of controls, suggesting that personal experience of food poisoning can reduce (but not eliminate) optimistic bias. The disclosure of severe consequences experienced by others can also trigger teachable moments, especially if the actors or victims involved bear similarity to your target audience. As such, high-profile events such as the reporting of a large number of cases of food-borne illness or widespread contamination or recall of product can (temporarily) focus people's attention on the problem, perhaps reduce their illusion of invulnerability, and therefore provide opportunities to get food safety messages across.

Using Fear Appeals To Influence Behavior Change

Some risk communicators attempt to trigger such opportunities themselves by creating "fear appeals": persuasive messages intended to motivate behavior change by illustrating consequences that arouse fear. In a fear appeal, the communicator essentially tells the audience: "If you don't do what I recommend, terrible consequences will result," or more commonly, "Look at these terrible consequences. They can happen to you. But, if you do what I recommend, you can avoid them." This persuasion strategy is very common, ranging from showing graphic movies to teenagers depicting the dire consequences of drunk driving to thirty-second toothpaste commercials warning of the "dangers of gingivitis." Indeed, advertisements based on fear appeals that warn of the dangers of "germs" and touting products to eliminate them abound (see "The 99.9% solution" above).

Note that a fear appeal message makes use of a problem-solution format. It identifies a potential fear-inducing problem and recommends a solution (O'Keefe, 2002a). In their review of more than 50 years of literature devoted to the effectiveness of fear appeals, Witte and Allen (2000) identify three key independent variables: fear, perceived threat, and perceived efficacy. Fear is defined as a highly arousing negative emotion. Perceived threat is a combination of the perceived severity of the threat and perceptions of one's susceptibility to it. Perceived efficacy is also composed of two components: perceived self-efficacy (the belief that one has the ability to perform the recommended or required actions) and perceived response efficacy (the belief that the action to be taken has an effect on reducing or eliminating the threat).

Some competing explanations for how fear appeals work include drive theories, parallel response models, and subjective expected utility models. (For discussion of some of these, see Dillard, 1994; Eagly and Chaiken, 1993; Witte, 1998; Witte and Allen, 2000; O'Keefe, 2002a). However, much of the disagreement centers on whether reducing the sense of fear or diminishing the perceived threat is the source of motivation of fear appeals. Yet, while fear is thought of as an emotion and perceived threat is typically characterized as cognition, they are integrally related; greater perceived threats seem to generate greater fear and high levels of fear appear to increase the assessment of threat. Moreover, several of the parallel response models suggest that people are motivated by trying to minimize both the threat and the resulting fear (Witte and Allen, 2000).

As such, in his summary of the research evidence O'Keefe (2002a) concludes that. In general, stronger fear appeal messages with more intense contents do arouse greater fear, though there are lots of variations in what individuals find fearful. So, influencing an audience's level of fear is not always easy, but greater induced fear seems to be more persuasive. Witte and Allen (2000) also conclude that stronger fear appeals can produce both greater fear and higher levels of perceived severity and susceptibility. As a result, the evidence shows that messages with stronger contents are more persuasive in changing attitudes, intentions, and actions, though this effect is not as great as one might hope (correlations range from 0.10 to 0.20). Finally, the stronger the perceived response efficacy and self-efficacy, the more likely people are to change their attitudes, intentions, and behaviors toward the recommended response.

They conclude that risk communicators can best develop effective fear appeal messages by increasing references to both the severity of the threat and the target audience's susceptibility to it. Messages that make a health issue seem both serious and likely are the most motivating. However, strong fear appeals only work when accompanied by equally strong efficacy messages that make people believe both that they can perform a recommended action and that the action will avert or minimize a threat. They warn that strong fear appeals should be used cautiously since they can backfire if the target audience does not believe that they can do something about the threat presented.

Using Other Emotions To Influence Behavior
Although fear appeals are the best-studied messages, persuasive messages built on other emotions such as pride, disgust, social stigma, shame, embarrassment, and especially guilt are also common (O'Keefe, 2000, 2002a, 2002b). Like fear appeals, they are intended to arouse an emotional state paired with a recommended action that provides a means to avoid those feelings. The success of products to combat "halitosis," "body odor," and "ring around the collar" is largely the result of advertising messages built on such appeals (see "How do people come to know about microbes?" above).

While appeals to emotions have been shown to be effective in persuading people to take specific advocated actions, it is also clear that *anticipated* emotions can also influence intentions and actions (O'Keefe, 2002a). For example, research has shown both that people will try to avoid actions that they anticipate would make them feel guilty and that anticipated levels of guilt can be manipulated (Birkimer et al., 1993; O'Keefe, 2002b; O'Keefe and Figgé, 1997, 1999; Lindsey, 2005). As a result, as has been demonstrated by countless religions and generations of mothers, it is possible to influence a range of behaviors by tying their prohibition to anticipated feelings of guilt, shame, or embarrassment.

Getting People To Behave Reflexively

Finally, while it is important to increase knowledge and provide cues to take appropriate actions to prevent microbial contamination, these are essentially strategies to get people to actively think about what they are doing at a particular moment and to modify their behaviors accordingly. Yet, often, the real goal is to get people to practice these behaviors reflexively, *without* having to think about them; that is, to turn them into habits.

This can typically only happen with repeated practice and in an environment that supports these behaviors both physically and culturally, that is, in places where they are easy to carry out and where they are expected as a norm. For example, in their study of nursing students, Snow et al. (2006) found that while the students had an overall low rate of hand hygiene, the strongest predictors of hand washing and glove usage were the hand hygiene practices of their mentors. Similarly, Muto (2000) found that placing alcohol-based hand antiseptic dispensers outside every door on two hospital wards did not increase the rate of compliance with hand hygiene practices. However, physician hand compliance *was* associated with the hygiene behaviors practiced by attending physicians whose example was usually followed by all other physicians on rounds. As such, getting influential people to model appropriate behaviors can be a powerful way to influence the actions of entire groups. Again, this is a lesson well learned by marketers of consumer products, who use familiar, respected spokespersons to endorse or model their wares.

CONCLUSION

It is clear that many people do not have a good understanding of microbiology, food-preparation hygiene, or food-borne diseases. They also overestimate their own control and efficacy in preventing microbial contamination and they underestimate the likelihood that their actions (or inactions) will result in contamination or food-borne illness. It is also apparent that participation in formal food safety training programs is not enough to ensure that even professional food handlers will practice proper procedures. So, what can be done?

First, recognize that because risk communication involves human beings, it is not an exact science and has its limitations. Advertisers, with millions of dollars in resources behind them, have learned that even award-winning marketing campaigns are rarely effective enough to influence even the minority of an audience. Also realize that because of the infinite variability of humans and the circumstances in which you find them, how people are liable to respond to what you have to say is also likely to vary substantially between audiences and over time. As a result, there is no guarantee that what proves effective in one situation will necessarily work in another. So, accept that your power to inform and persuade is fundamentally limited even under the best circumstances. However, that does not mean you should give up. It just means that you need to be realistic about what you expect to accomplish.

As detailed in this chapter, your odds of success are improved if you start out with clear goals for communicating. It is also important to thoroughly consider the needs and abilities of each audience with which you intend to interact. This will require assessing what people already know and believe about the risk issues you want to talk about and recognizing the potential barriers you may face in getting your message across, including varying levels of illiteracy and innumeracy. Constructing appropriate messages will also require an appreciation of how people perceive risks, in general, and microbial risks more specifically, as well as a good understanding of the cultural and psychological influences that help to determine those perceptions. Appropriately targeting those messages will also depend a great deal on finding suitable channels that are likely to reach your audience at the right time and in the right place. Finally, changing behaviors will require an awareness of how people *currently* behave toward particular microbial risks and an appreciation of the psychological, sociological, anthropological, and economic factors that drive and maintain those behaviors.

A Final Thought—It Ain't Rocket Science

A few years ago, a reporter from National Public Radio wondered about the phrase "it isn't rocket science" as a comparator used to describe something that was not very difficult. In an effort to find out how difficult rocket science actually is, the reporter interviewed an aeronautical engineer. The engineer explained that while rocket science could indeed be complicated, the calculations necessary are based on formulas derived from well-known principles of physics. "It isn't brain surgery," he replied.

So, the reporter interviewed a neurosurgeon to find out how difficult surgery on the brain really is. The surgeon responded that brain surgery obviously required good eye-hand coordination and an excellent understanding of the anatomy of the brain. However, with some variations, the anatomical structures of the brain are pretty much the same from person to person. "It's not like its psychology," the neurosurgeon replied. "Since every human is different, there are few things more difficult than trying to change people's behavior."

That's microbial risk communication. It isn't rocket science. It isn't brain surgery. It's harder. Get started.

REFERENCES

A. Wulfing & Company. If you could see the germs in your throat. Medicine and Madison Avenue On-Line Project—Document MM0025, John W. Hartman Center for Sales, Advertising & Marketing History, Rare Book, Manuscript, and Special Collections Library. Duke University, Durham, NC. [Online.] http://scriptorium.lib.duke.edu/mma/. Accessed 30 October 2005.

Ackerley, L. 1994. Consumer awareness of food hygiene and food poisoning. *Environ. Health* 102:69–74.

Altekruse, S. F., D. A. Street, S. B. Fein, and A. S. Levy. 1996. Consumer knowledge of foodborne microbial hazards and food handling practices. *J. Food Prot.* 59:287–294.

Altekruse, S. F., and D. L. Swerdlow. 1996. The changing epidemiology of foodborne diseases. *Am. J. Med. Sci.* 311(1):23–29.

Altekruse, S. F., S. Yang, B. B. Timbo, and F. J. Angulo. 1999. A multi-state survey of consumer food-handling and food consumption practices. *Am. J. Prev. Med.* 16(3):216–221.

American Society for Microbiology. 2003. Another US airport travel hazard—dirty hands. [Online.] http://www.eurekalert.org/pub_releases/2003-09/asfm-aua091103.php. Accessed 29 January 2006.

American Society for Microbiology. 2005. Women better at hand hygiene habits, hands down. [Online.] http://www.asm.org/Media/index.asp?bid=38075. Accessed 29 January 2006.

Angyal, A. 1941. Disgust and related aversions. *J. Abnorm. Soc. Psychol.* 36:393–412.

Barnes, J. (ed.). 1984. *The Complete Works of Aristotle: the Revised Oxford Translation*. Princeton University Press, Princeton, NJ.

Birkimer, J. C., P. L. Johnston, and M. M. Berry. 1993. Guilt and help from friends: variables related to healthy behavior. *J. Soc. Psychol.* 133:683–692.

Bittar, C. 1999. Lysol, Clorox, spray invective on disinfectants. *Brandweek* 40(40):4.

Brandt, M. B., C. J. Spease, G. June, and A.-M. Brown. 2003. Prevalence of food safety, quality, and other consumer statements on labels of processed, packaged foods. *Food Prot. Trends* 23(11):870–881.

Brewer, M. S., G. K. Sprouls, and C. Russon. 1994. Consumer attitudes towards food safety issues. *J. Food Saf.* 14:63–76.

Bynum, C. W. 1985. Fast, feast, & flesh: the religious significance of food to medieval women. *Representations* **11**:1–25.

Clayton, D. A., C. J. Griffith, and P. Price. 2003. An investigation of the factors underlaying consumers' implementation of specific food safety practices. *Br Food J.* **105**(7):434–453.

Clayton, D. A., C. J. Griffith, P. Price, and A. C. Peters. 2002. Food handlers' beliefs and self-reported practices. *Int. J. Environ. Health Res.* **12**(1):25–39.

Clorox. November 2005a. Wipe out winter woes. *Good Housekeeping* Advertising Insert at p. 236.

Clorox. 2005b. Germs 101: learn about germs. [Online.] http://www.cloroxclassrooms.com/learnAboutGerms.html. Accessed 31 October 2005.

Coleman, J. S. 1990. *Foundations of Social Theory*. Harvard University Press, Cambridge, MA.

Coleman, P., and C. Griffith. 1998. Risk assessment: a diagnostic tool for caterers. *Hospitality Manag.* **17**:289–301.

Covello, V. T. 1992. Trust and credibility in risk communication. *Health Environ. Dig.* **6**(1):1–3.

Covello, V. T. 1993. Risk communication and occupational medicine. *J. Occup. Med.* **35**(1):18–19.

Curtis, V., and A. Biran. 2001. Dirt, disgust, and disease: is hygiene in our genes? *Perspect. Biol. Med.* **44**(1):17–31.

Curtis, V., R. Aunger, and T. Rabie. 2004. Evidence that disgust evolved to protect from risk of disease. *Proc. Biol. Sci. R. Soc. Lond.* **271**:S131–S133.

Daniels, R. W. 1998. Home food safety. *Food Technol.* **52**:1405–1411.

Darwin, C. 1872. *The Expression of the Emotions in Man and Animals*, 1965 ed. University of Chicago Press, Chicago, IL.

Davies, S., H. Haines, B. Norris, and J. R. Wilson. 1998. Safety pictograms: are they getting the message across? *Appl. Ergon.* **29**(1):15–23.

Dillard, J. P. 1994. Rethinking the study of fear appeals: an emotional perspective. *Commun. Theory* **4**:295–323.

Douglas, M. 1966. *Purity and Danger*. Routledge, London, United Kingdom.

Douglas, M. Winter 1972. Deciphering a meal. *Daedalus* **10**:61–81.

Douglas, M., and A. Wildavsky. 1982. *Risk and Culture: an Essay on the Selection of Technical and Environmental Dangers*. University of California Press, Berkeley, CA.

Drackett Company. 1949. If you could swab out your sink drain . . . Medicine and Madison Avenue On-Line Project—Document MM0777, John W. Hartman Center for Sales, Advertising & Marketing History, Rare Book, Manuscript, and Special Collections Library. Duke University, Durham, NC. [Online.] http://scriptorium.lib.duke.edu/mma/. Accessed 30 October 2005.

Eagly, A. H., and S. Chaiken. 1993. *The Psychology of Attitudes*. Harcourt Brace Jovanovich, Fort Worth, TX.

Einsiedel, E. F. 2000. Understanding 'publics' in the public understanding of science, p. 205–216. *In* M. Dierkes and C. von Grote (ed.), *Between Understanding and Trust: the Public, Science and Technology*. Harwood Academic Publishers, Amsterdam, The Netherlands.

Fallon, A. E., and P. Rozin. 1983. The psychological bases of food rejections by humans. *Ecol. Food Nutr.* 13:5–26.

Fallon, A. E., P. Rozin, and P. Pliner. 1984. The child's conception of food: the development of food rejections with special reference to disgust and contamination sensitivity. *Child Dev.* 55:566–575.

Fein, S. B., C.-T. J. Lin, and A. S. Levy. 1995. Foodborne illness: perceptions, experience, and preventive behaviors in the United States. *J. Food Prot.* 58(12):1405–1411.

Fiddes, N. 1994. Social aspects of meat eating. *Proc. Nutr. Soc.* 53:271–280.

Fife-Shaw, C., and G. Rowe. 1996. Public perceptions of everyday food hazards: a psychometric study. *Risk Anal.* 16(4):487–500.

Finch, J. E., J. Prince, and M. Hawksworth. 1978. A bacteriological survey of the domestic environment. *J. Appl. Bacteriol.* 45:357–364.

Fischhoff, B., P. Slovic, S. Lichtenstein, S. Read, and B. Combs. 1978. How safe is safe enough? A psychometric study of attitudes towards technological risks and benefits. *Policy Sci.* 9:127–152.

Food and Drink Federation (FDF) and the Institute of Environmental Health Officers (IEHO). 1995. *National Food Safety Report.* FDF-IEHO, London, United Kingdom.

Foodlink. 2005. Press notice: Dramatic drop in food poisoning cases—18% decrease since campaign focused on hand washing. [Online.] http://www.foodlink.org.uk/press_releases/pressnational_050613_foodlink.pdf. Accessed 29 January 2006.

Food Marketing Institute. 2002. *Trends in the United States: Consumer Attitudes & the Supermarket, 2002.* Food Marketing Institute, Washington, DC.

Food Standards Agency. 2002. Catering Workers Hygiene Survey 2002. [Online.] http://www.food.gov.uk/wales/pressreleases/2002/oct/97264. Accessed 29 January 2006.

Food Standards Agency. 2003. Consumer attitudes to food standards—UK. [Online.] http://www.food.gov.uk/multimedia/pdfs/cas2002uk.pdf. Accessed 29 January 2006.

Frazer, J. G. 1890. *The New Golden Bough: a Study in Magic and Religion,* 1959 abridged ed., T. H. Gaster (ed.). Macmillan, New York, NY.

Frewer, L. J., C. Howard, D. Hedderley, and R. Shepherd. 1995. What determines trust in information about food-related risks? Underlying psychological constructs. *Risk Anal.* 15(4):473–486.

Frewer, L. J., C. Howard, D. Hedderley, and R. Shepherd. 1997. The elaboration likelihood model and communication about food risks. *Risk Anal.* 17(6):759–770.

Frewer, L. J., C. Howard, and R. Shepherd. 1997. Public concerns in the United Kingdom about general and specific applications of genetic engineering: risk, benefit and ethics. *Sci. Technol. Hum. Values* 22(1):98–124.

Frewer, L. J., R. Shepherd, and P. Sparks. 1994. The interrelationship between perceived knowledge control and risk associated with a range of food-related hazards targeted at the individual, other people and society. *J. Food Saf.* 14:19–40.

Glassner, B. 1999. *The Culture of Fear: Why Americans Are Afraid of the Wrong Things.* Basic Books, New York, NY.

Griffith, C. J., K. A. Mathias, and P. E. Price. 1994. The mass media and food hygiene education. *Br. Food J.* **96**(9):16–21.

Goleman, D. 1995. *Emotional Intelligence: Why It Can Matter More Than IQ.* Bantam Books, New York, NY.

Gray, G. M., and D. P. Ropeik. 2002. Dealing with the dangers of fear: the role of risk communication. *Health Affairs* **21**(6):106–116.

Green, L., C. Selman, A. Banerjee, R. Marcus, C. Medus, F. J. Angulo, V. Radke, S. Buchanan, and EHS-Net Working Group. 2005. Food service workers' self-reported food preparation practices: an EHS-Net study. *Int. J. Hyg. Environ. Health* **208**:27–35.

Griffith, C. J., K. A. Mathias, and P. E. Price. 1994. The mass media and food hygiene education. *Br. Food J.* **96**(9):16–21.

Griffith, C. J., D. Worsfold, and R. Mitchell. 1998. Food preparation, risk communication and the consumer. *Food Control* **9**(4):225–232.

Haddix, D. 1990. Alarm as a media event. *Columbia J. Rev.* **28**:44–45.

Hallman, W. K., and S. C. Condry. 2005. Communicating about microbial risks. Paper presented to the 17th Annual Conference of the International Society for Environmental Epidemiology (ISEE). Johannesburg, South Africa.

Hance, B. J., C. Chess, and P. M. Sandman. 1988. *Improving Dialogue with Communities: a Manual for Government.* Department of Environmental Protection, Trenton, NJ.

Hansen, J., L. Holm, L. Frewer, P. Robinson, and P. Sandøe. 2003. Beyond the knowledge deficit: recent research into lay and expert attitudes to food risks. *Appetite* **41**:111–121.

Harcar, T., and F. Karakaya. 2005. A cross-cultural exploration of attitudes toward product expiration dates. *Psychol. Mark.* **22**(4):353–371.

Hardin, R. 2002. *Trust and Trustworthiness.* Russell Sage Foundation, NewYork, NY.

Harris, M. 1985. *Good to Eat.* Simon and Schuster, New York, NY.

Harrison, K. 2001. Too close to home: dioxin contamination of breast milk and the political agenda. *Policy Sci.* **34**:35–62.

Hayes, D. J., J. F. Shogren, S. Y. Shin, and J. B. Kliebenstein. 1995. Valuing food safety in experimental auction markets. *Am. J. Agric. Econ.* **77**:40–53.

Haysom, L., and K. Sharp. 2004. Cross-contamination from raw chicken during meal preparation. *Br. Food J.* **106**(1):38–50.

Howes, M., S. McEwen, M. Griffiths, and L. Harris. 1996. Food handler certification by home study: measuring changes in knowledge and behavior. *Dairy Food Environ. Sanit.* **16**:737–744.

Hunter, B. T. 2000. Food for thought: germ-proofed products? *Consumer's Res.* **83**(4):8–9.

Hynson, Westcott and Dunning, Inc. 1946. For Their Protection Mercurochrome. Medicine and Madison Avenue On-Line Project—Document MM0189, John W. Hartman Center for Sales, Advertising & Marketing History, Rare Book, Manuscript, and Special Collections Library. Duke University, Durham, NC. [Online.] http://scriptorium.lib.duke.edu/mma/. Accessed 30 October 2005.

JAMA. 1931. Listerine and other mouth washes. *JAMA* **96**(16):1308–1309.

Jasanoff, S. 2000. The 'science wars' and American politics, p. 39–60. *In* M. Dierkes and C. von Grote (ed.), *Between Understanding and Trust: the Public, Science and Technology*. Harwood Academic Publishers, Amsterdam, The Netherlands.

Jay, J. S., D. Comar, and L. D. Govenlock. 1999. A video study of Australian domestic food-handling practices. *J. Food Prot.* **63**:1285–1296.

Johnson, A. E., A. J. M. Donkin, K. Morgan, J. M. Lilley, R. J. Neale, R. M. Page, and R. Silburn. 1998. Food safety knowledge and practice among elderly people living at home. *J. Epidemiol. Community Health* **52**:745–748.

Johnson, B. B. 1999. Exploring dimensionality in the origins of hazard-related trust. *J. Risk Res.* **2**:325–354.

Johnson & Johnson. 1943. An old family friend in a new form—Sulfa-thiazole Band-Aid! Medicine and Madison Avenue On-Line Project—Document MM0169, John W. Hartman Center for Sales, Advertising & Marketing History, Rare Book, Manuscript, and Special Collections Library. Duke University, Durham, NC. [Online.] http://scriptorium.lib.duke.edu/mma/. 30 Accessed October 2005.

Kasperson, R. E. 1986. Six propositions on public participation and their relevance for risk communication. *Risk Anal.* **6**(3):275–281.

Kasperson, R. E., D. Golding, and S. Tuler. 1992. Social distrust as a factor in siting hazardous facilities and communicating risks. *J. Soc. Issues* **48**(4):161–187.

Kendall, P., L. C. Medeiros, V. Hillers, G. Chen, and S. DiMascola. 2003. Food handling behaviors of special importance for pregnant women, infants and young children, the elderly, and immune-compromised people. *J. Am. Diet. Assoc.* **103**:1646–1649.

Kimberly-Clark. November 2005. Ruthless killer. *Good Housekeeping*, p. 67.

Klontz, K. C., B. Timbo, S. Fein, and A. Levy. 1995. Prevalence of selected food consumption and preparation behaviors associated with foodborne disease. *J. Food Prot.* **58**:927–930.

Lambert Pharmacal Company. 1917. Listerine—the safe antiseptic. Medicine and Madison Avenue On-Line Project—Document MM0600, John W. Hartman Center for Sales, Advertising & Marketing History, Rare Book, Manuscript, and Special Collections Library. Duke University, Durham, NC. [Online.] http://scriptorium.lib.duke.edu/mma/. Accessed 30 October 2005.

Lambert Pharmacal Company. 1929a. What I know about *nice* women. Medicine and Madison Avenue On-Line Project—Document MM0612, John W. Hartman Center for Sales, Advertising & Marketing History, Rare Book, Manuscript, and Special Collections Library. Duke University, Durham, NC. [Online.] http://scriptorium.lib.duke.edu/mma/. Accessed 30 October 2005.

Lambert Pharmacal Company. 1929b. Street car colds! Protect yourself with Listerine. Kills germs in 15 seconds. Medicine and Madison Avenue On-Line Project—Document MM0612, John W. Hartman Center for Sales, Advertising & Marketing History, Rare Book, Manuscript, and Special Collections Library. Duke University, Durham, NC. [Online.] http://scriptorium.lib.duke.edu/mma/. Accessed 30 October 2005.

Lambert Pharmacal Company. 1930a. Wet feet breed sore throat. Medicine and Madison Avenue On-Line Project—Document MM0626, John W. Hartman Center for Sales, Advertising & Marketing History, Rare Book, Manuscript, and Special Collections Library. Duke

University, Durham, NC. [Online.] http://scriptorium.lib.duke.edu/mma/. Accessed 30 October 2005.

Lambert Pharmacal Company. 1930b. O careful mother before baby's meals rid your hands of germs. Medicine and Madison Avenue On-Line Project—Document MM0645, John W. Hartman Center for Sales, Advertising & Marketing History, Rare Book, Manuscript, and Special Collections Library. Duke University, Durham, NC. [Online.] http://scriptorium.lib.duke.edu/mma/. Accessed 30 October 2005.

Lambert Pharmacal Company. 1932. Is your child in school? Look out for ringworm. Medicine and Madison Avenue On-Line Project—Document MM0658, John W. Hartman Center for Sales, Advertising & Marketing History, Rare Book, Manuscript, and Special Collections Library. Duke University, Durham, NC. [Online.] http://scriptorium.lib.duke.edu/mma/. Accessed 30 October 2005.

Lambert Pharmacal Company. 1934. Their spirit lives on. Medicine and Madison Avenue On-Line Project—Document MM0683, John W. Hartman Center for Sales, Advertising & Marketing History, Rare Book, Manuscript, and Special Collections Library. Duke University, Durham, NC. [Online.] http://scriptorium.lib.duke.edu/mma/. Accessed 30 October 2005.

Lambert Pharmacal Company. 1937. The most important announcement we have ever made. Dandruff can be cured with Listerine! Medicine and Madison Avenue On-Line Project—Document MM0690, John W. Hartman Center for Sales, Advertising & Marketing History, Rare Book, Manuscript, and Special Collections Library. Duke University, Durham, NC. [Online.] http://scriptorium.lib.duke.edu/mma/. Accessed 30 October 2005.

Lambert Pharmacal Company. 1946. The job hunter with two strikes against him. Medicine and Madison Avenue On-Line Project—Document MM0185, John W. Hartman Center for Sales, Advertising & Marketing History, Rare Book, Manuscript, and Special Collections Library. Duke University, Durham, NC. [Online.] http://scriptorium.lib.duke.edu/mma/. Accessed 30 October 2005.

Lambert Pharmacal Company. 1956. Often a bridesmaid... never a bride! Medicine and Madison Avenue On-Line Project—Document MM0700, John W. Hartman Center for Sales, Advertising & Marketing History, Rare Book, Manuscript, and Special Collections Library. Duke University, Durham, NC. [Online.] http://scriptorium.lib.duke.edu/mma/. Accessed 30 October 2005.

Lambert Pharmacal Company. 1958. No matter where you go now . . . you can't escape from germs! Medicine and Madison Avenue On-Line Project—Document MM0710, John W. Hartman Center for Sales, Advertising & Marketing History, Rare Book, Manuscript, and Special Collections Library, Duke University. Durham, NC [Online.] http://scriptorium.lib.duke.edu/mma/. Accessed 30 October 2005.

Lambert Pharmacal Company. 1959. Wherever children gather these days, you're sure to find germs! Medicine and Madison Avenue On-Line Project—Document MM0715, John W. Hartman Center for Sales, Advertising & Marketing History, Rare Book, Manuscript, and Special Collections Library. Duke University, Durham, NC. [Online.] http://scriptorium.lib.duke.edu/mma/. Accessed 30 October 2005.

Lang, J. T., and W. K. Hallman. 2005. Who does the public trust? The case of genetically modified food in the United States. *Risk Anal.* 25(5):1241–1252.

Leman, P. 2001. Clinical and microbiological features of suspect sporadic food poisoning cases presenting to an accident and emergency department. *Commun. Dis. Public Health.* **4:**209–212.

Lesch, M. F. 2003. Comprehension and memory for warning symbols: age-related differences and impact of training. *J. Saf. Res.* **34:**495–505.

Lever Brothers Ltd. 1938. No ordinary soap stops "B.O." as Lifebuoy does! Medicine and Madison Avenue On-Line Project—Document MM0295, John W. Hartman Center for Sales, Advertising & Marketing History, Rare Book, Manuscript, and Special Collections Library. Duke University, Durham, NC. [Online.] http://scriptorium.lib.duke.edu/mma/. Accessed 30 October 2005.

Lever Brothers Ltd. 1939. Mothers! I've discovered a way to get kiddies to wash hands and like it. Medicine and Madison Avenue On-Line Project—Document MM0292, John W. Hartman Center for Sales, Advertising & Marketing History, Rare Book, Manuscript, and Special Collections Library. Duke University, Durham, NC. [Online.] http://scriptorium.lib.duke.edu/mma/. Accessed 30 October 2005.

Lévi-Strauss, C. 1966. The culinary triangle. *New Soc.* **166:**937–940.

Lévi-Strauss, C. 1970. *The Raw and the Cooked.* Cape, London, United Kingdom.

Levitt, S. D., and S. J. Dubner. 2005. *Freakonomics: a Rogue Economist Explores the Hidden Side of Everything.* HarperCollins Publishers Inc., New York, NY.

Lin, C.-T. J., K. L. Jensen, and S. T. Yen. 2005. Awareness of foodborne pathogens among US consumers. *Food Qual. Pref.* **16:**401–412

Lindsey, L. L. M. 2005. Anticipated guilt as behavioral motivation: an examination of appeals to help unknown others through bone marrow donation. *Hum. Commun. Res.* **31(4):**453–481.

Marchand, R. 1985. *Advertising the American Dream: Making Way for Modernity, 1920–1940.* University of California Press, Berkeley, CA.

Martins, Y., and P. Pliner. 2006. "Ugh! That's disgusting!": identification of the characteristics of foods underlying rejections based on disgust. *Appetite* **46:**75–85.

Mauss, M. 1902. *A General Theory of Magic* (R. Brain, trans.), 1972 ed. W. W. Norton, New York, NY.

McCarthy, M., M. Brennan, A. L. Kelly, C. Ritson, M. de Boer, and N. Thompson. 2007/3. Who is at risk and what do they know? Segmenting a population on their food safety knowledge. *Food Qual. Pref.* **18:**205–217.

McGinnies, E., and C. D. Ward. 1980. Better liked than right: trustworthiness and expertise as factors in credibility. *Pers. Soc.Psychol. Bull.* **6(3):**467–472.

McKie, L. J., and R. Wood. 1991 People's source of recipes: some implications for an understanding of food related behaviour. *Br. Food J.* **94:**12–17.

Meredith, L., R. Lewis, and M. Haslum. 2001. Contributory factors to the spread of contamination in a model kitchen. *Br. Food J.* **103(1):**23–35.

Michigan Chemical Company. 1950. Flies are a serious problem. Medicine and Madison Avenue On-Line Project—Document MM0812, John W. Hartman Center for Sales, Advertising & Marketing History, Rare Book, Manuscript, and Special Collections Library. Duke University, Durham, NC. [Online.] http://scriptorium.lib.duke.edu/mma/. Accessed 30, October 2005.

Miles, S., D. S. Braxton, and L. J. Frewer. 1999. Public perceptions about microbiological hazards in food. *Br. Food J.* **101**(10):744–762.

Miles, S., and L. J. Frewer. 2001. Investigating specific concerns about different food hazards. *Food Qual. Pref.* **12**:47–61.

Miles, S., and V. Scaife. 2003. Optimistic bias and food. *Nutr. Res. Rev.* **16**(1):3–19.

Miller, L., P. Rozin, and A. P. Fiske. 1998. Food sharing and feeding another person suggest intimacy; two studies of American college students. *Eur. J. Soc. Psychol.* **28**:423–436.

Moeller, S. 1999. *Compassion Fatigue: How the Media Sell Disease, War, Famine, and Death.* Routledge, New York, NY.

Moore, J., A. Earless, and T. Parsons. 1992. Women's magazines: their influence on nutritional knowledge and food habits. *Nutr. Food Sci.* **92**(3):18–21.

Morgan, M. G., B. Fischoff, A. Bostrom, and C. Atman. 2002. *Risk Communication: a Mental Models Approach.* Cambridge University Press, New York, NY.

Mortlock, M. P., A. C. Peters, and C. J. Griffith. 1999. Food hygiene and hazard analysis critical control point in the United Kingdom food industry: practices, perceptions and attitudes. *J. Food Prot.* **62**:786–792.

Moxley, L., and A. Sanford. 1993. *Communicating Quantities: a Psychological Perspective.* Erlbaum, Hillsdale, NJ.

Muto, C. A. 2000. Hand hygiene rates unaffected by installation of dispensers of a rapidly acting hand antiseptic. *Am. J. Infect. Control.* **28**(3):273–276.

National Research Council. 1989. *Improving Risk Communication.* National Academies Press, Washington, DC.

National Science Board. 2002. *Science and Engineering Indicators—2002,* NSB-02-1. National Science Foundation, Arlington, VA.

National Science Board. 2006. *Science and Engineering Indicators 2006,* 2 vols. (vol. 1, NSB 06-01; vol. 2, NSB 06-01A). National Science Foundation, Arlington, VA.

National Virtual Translation Center. 2006. Languages spoken in the U.S. [Online.] http://www.nvtc.gov/lotw/months/november/USlanguages.html. Accessed 14 September 2006.

Nucci, M. L., J. DaLoia, V. Frenkel, and W. K. Hallman. 2006. Prime time food, or, what can you learn about food safety on the evening news? Paper presented at the Joint 2006 Annual Meetings of the Association for the Study of Food and Society (ASFS) and the Agriculture, Food, and Human Values Society (AFHVS), Boston, MA.

O'Boyle, C. A., S. J. Henly, and E. Larson. 2001. Understanding adherence to hand hygiene recommendations: the theory of planned behavior. *Am. J. Infect. Control* **29**(6):352–360.

O'Keefe, D. J. 2000. Guilt and social influence, p. 67–101. *In* M. E. Roloff (ed.), *Communication Yearbook 23.* Sage, Thousand Oaks, CA.

O'Keefe, D. J. 2002a. *Persuasion: Theory & Research,* 2nd ed. Sage, Thousand Oaks, CA.

O'Keefe, D. J. 2002b. Guilt as a mechanism of persuasion, p. 329–344. *In* J. P. Dillard and M. Pfau (ed.), *The Persuasion Handbook: Developments in Theory and Practice.* Sage, Thousand Oaks, CA.

O'Keefe, D. J., and M. Figgé. 1997. A guilt-based explanation of the door-in-the-face influence strategy. *Hum. Commun. Res.* **24**:64–81.

O'Keefe, D. J., and M. Figgé. 1999. Guilt and expected guilt in the door-in-the-face technique. *Commun. Monogr.* **66**:312–324.

Palmer, S., H. Houston, B. Lervy, D. Riberio, and P. Thomas. 1996. Problems in the diagnosis of foodborne infection in general practice. *Epidemiol. Infect.* **117**(3):479–484.

Parry, S. M., S. Miles, T. Ascanio, S. R. Palmer, and South and East Wales Infectious Disease Group. 2004. Differences in perception of risk between people who have and have not experienced *Salmonella* food poisoning. *Risk Anal.* **24**(1):289–299.

Patil, S. R., R. Morales, S. Cates, D. Anderson, and D. Kendall. 2004. An application of meta-analysis in food safety consumer research to evaluate consumer behaviors and practices. *J. Food Prot.* **67**(11):2587–2595.

Paulos, J. A. 1988. *Innumeracy: Mathematical Illiteracy and Its Consequences.* Hill & Wang, New York, NY.

Peters, H. P. 2000. From information to attitudes? Thoughts on the relationship between knowledge about science and technology and attitudes toward technologies, p. 265–286. *In* M. Dierkes and C. von Grote (ed.), *Between Understanding and Trust: the Public, Science and Technology.* Harwood Academic Publishers, Amsterdam, The Netherlands.

Peters, R. G., V. T. Covello, and D. B. McCallum. 1997. The determinants of trust and credibility in environmental risk communication: an empirical study. *Risk Anal.* **17**(1):43–54.

Petty, R. E., and J. T. Cacioppo. 1986. *Communication and Persuasion: Central and Peripheral Routes to Attitude Change.* Springer-Verlag, New York, NY.

Pfizer. 2005a. History of Listerine. [Online.] http://www.pfizer.com.au/Products/Listerine/History.aspx. Accessed 20 November 2005.

Pfizer. 2005b. You can touch, then use Purell®. [Online.] www.purell.com. Accessed 31 October 2005.

Poortinga, W., and N. Pidgeon. 2003. Exploring the dimensionality of trust in risk regulation. *Risk Anal.* **23**(5):961–972.

Powell, S. C., R. W. Attwell, and S. J. Massey. 1997. The impact of training on knowledge and standards of food hygiene—a pilot study. *Int. J. Environ. Health Res.* **7**:329–334.

Procter and Gamble. 2006. Crest Pro-health Rinse. [Online.] http://www.crest.com/prohealthrinse/index.jsp. Accessed 19 July 2006.

Reckitt Benckiser. 2005. LYSOL® Sanitizing Wipes. [Online.] http://www.lysol.com/solutionsfinder_sanwipes.shtml. Accessed 30 October 2005.

Redmond, E. C., and C. J. Griffith. 2003. A comparison and evaluation of research methods used in consumer food safety studies. *Int. J. Consum. Stud.* **27**(1):17–33.

Redmond, E. C., and C. J. Griffith. 2004. Consumer perceptions of food safety risk, control and responsibility. *Appetite* **43**:309–313

Redmond, E. C., and C. J. Griffith. 2005a. Consumer perceptions of food safety education sources: implications for effective strategy development. *Br. Food J.* **107**(7):467–483.

Redmond, E. C., and C. J. Griffith. 2005b. Factors influencing the efficacy of consumer food safety communication. *Br. Food J.* **107**(7):484–499.

Redmond, E. C., C. J. Griffith, J. Slader, and T. J. Humphrey. 2004. Microbiological and observational analysis of cross contamination risks during domestic food preparation. *Br. Food J.* **106**(8):581–597.

Reid, A., D. Wood, and D. Kinney. 1998. Food hygiene information: power to the people? *Nutr. Food Sci.* **3**:138–144.

Renn, O., and D. Levine. 1991. Credibility and trust in risk communication. *In* R. E. Kasperson and P. J. Stallen (ed.), *Communicating Risks to the Public*. Kluwer Academic Publishers, Dordrecht, The Netherlands.

Rennie, D. M. 1994. Evaluation of food hygiene education. *Br. Food J.* **11**:20 25.

Rother, H.-A. March 2006. Risk perception, risk communication, and the effectiveness of pesticide labels in communicating hazards to South African farm workers. Dissertation Abstracts International, DAI-A 66/09, p. 3481.

Rotter, J. B. 1971. Generalized expectancies for interpersonal trust. *Am. Psychol.* **26**:443–452.

Rozin, P., J. Haidt, and C. McCauley. 2000. Disgust, p. 637–653. *In* M. Lewis and J. M. Haviland (ed.), *Handbook of Emotions*, 2nd ed. Guilford Press, New York, NY.

Rozin, P., and A. E. Fallon. 1980. The psychological categorization of foods and non foods: a preliminary taxonomy of food rejections. *Appetite* **1**:193–201.

Rozin, P., and A. E. Fallon. 1987. A perspective on disgust. *Psychol. Rev.* **94**(1):23–41.

Rozin, P., A. E. Fallon, and R. Mandell. 1984. Family resemblance in attitudes to foods. *Dev. Psychol.* **21**:1075–1079.

Rozin, P., M. Markwith, and C. Nemeroff. 1992. Magical contagion beliefs and fear of AIDS. *J. Appl. Soc. Psychol.* **22**(14):1081–1092.

Rozin, E., L. Millman, and C. Nemeroff. 1986. Operation of the laws of sympathetic magic in disgust and other domains. *J. Pers. Soc. Psychol.* **50**:703–712.

Sadalla, E. K., and W. J. Burroughs. 1981. Profiles in eating. *Psychol. Today* **15**(10):51–57.

SC Johnson Company. 2005. Raid® with GermFighter® Ant & Roach Killer. [Online.] http://www.killsbugsdead.com/fop_ark_germ.asp. Accessed 30 October 2005.

Scott, E., and S. F. Bloomfield. 1990. The survival and transfer of microbial contamination via cloths, hands and utensils. *J. Appl. Bacteriol.* **68**:271–278.

Scott, E., and S. F. Bloomfield. 1993. An in-use study of the relationship between bacterial contamination of food preparation surfaces and cleaning cloths. *Lett. Appl. Microbiol.* **16**:173–177.

Scott, E., S. F. Bloomfield, and C. G. Barlow. 1982. An investigation of microbial contamination in the home. *J. Hyg.* **89**:279–293.

Scott Paper Company. 1959. See how Scotties keep cold germs from spreading. Medicine and Madison Avenue On-Line Project—Document MM0585, John W. Hartman Center for Sales, Advertising & Marketing History, Rare Book, Manuscript, and Special Collections Library. Duke University, Durham, NC. [Online.] http://scriptorium.lib.duke.edu/mma/. Accessed 30 October 2005.

Sharp, K., L. Haysom, and R. Parkinson. 2001. Anti-microbial hand washes for domestic use—their effectiveness in vitro and in normal use. *Int. J. Consum. Stud.* **25**(3):200–207.

Sharp, K., and R. P. Parkinson. 1999. A comparison of the growth of gram positive and gram negative food-poisoning bacteria on a variety of domestic dish cloths. *Consumer Sci. Res.* **18**(1):21–24.

Sharp, K., and H. Walker. 2003. A microbiological survey of communal kitchens used by undergraduate students. *Int. J. Consumer Stud.* **27**(1):11–16.

Slovic, P. 1987. Perception of risk. *Science* **236**:280–285.

Slovic, P. 2000. *The Perception of Risk*. Earthscan Publications, London, United Kingdom.

Slovic, P., M. L. Finucane, E. Peters, and D. G. MacGregor. 2004. Risk as analysis and risk as feelings: some thoughts about affect, reason, risk and rationality. *Risk Anal.* **24**: 311–322.

Slovic, P., B. Fischhoff, and S. Lichtenstein. 1980. Facts and fears: understanding perceived risk, p. 181–214. *In* R. C. Schwing and W. A. Albers, Jr. (ed.), *Societal Risk Assessment: How Safe Is Enough?* Plenum, New York, NY.

Slovic, P., B. Fischhoff, and S. Lichtenstein. 1985. Characterizing perceived risk, p. 91–123. *In* R. W. Kates, C. Hohenemser, and J. X. Kasperson (ed.), *Perilous Progress: Technology as Hazard*. Westview, Boulder, CO.

Slovic, P., B. Fischhoff, and S. Lichtenstein. 1986. The psychometric study of risk perception, p. 3–24. *In* V. T. Covello, J. Menkes, and J. Mumpower (ed.), *Risk Evaluation and Management*. Plenum Press, New York, NY.

Slovic, P., S. Lichtenstein, and B. Fischhoff. 1979. Images of disaster: perception and acceptance of risks from nuclear power, p. 223–245. *In* G. Goodman and W. Rowe (ed.), *Energy Risk Assessment*. Academic, London, United Kingdom.

Slovic, P., and E. Peters. 1998. The importance of worldviews in risk perception. *J. Risk Decis. Policy* **3**(2):165–170.

Snow, M., G. L. White, S. C. Alder, and J. B. Stanford. 2006. Mentor's hand hygiene practices influence student's hand hygiene rates. *Am. J. Infect. Control* **34**(1):18–24.

Sparks, P., and R. Shepherd. 1994a. Public perceptions of the potential hazards associated with food production and food consumption: an empirical study. *Risk Anal.* **14**(5):799–806.

Sparks, P., and R. Shepherd. 1994b. Public perceptions of food-related hazards: individual and social dimensions. *Food Qual. Pref.* **5**:185–194.

Speirs, J. P., A. Anderton, and J. H. Anderson. 1995. A study of the microbial content of the domestic kitchen. *Int. J. Environ. Health Res.* **5**:109–122.

Spriegel, G. 1991. Food safety in the home. *Nutr. Food Sci.* **91**:14–15.

Stearns. 1942. Help defend against disease by killing rats, mice and roaches. Medicine and Madison Avenue On-Line Project—Document MM0847, John W. Hartman Center for Sales, Advertising & Marketing History, Rare Book, Manuscript, and Special Collections Library. Duke University, Durham, NC. [Online.] http://scriptorium.lib.duke.edu/mma/. Accessed 30 October 2005.

Sturgis, P., and N. Allum. 2004. Science in society: re-evaluating the deficit model of public attitudes. *Public Understand. Sci.* **13**:55–74.

Taillard, M. 2000. Persuasive communication: the case of marketing. *UCL Working Papers in Linguistics* **12**:145–174.

The Dial Corporation. 2005. Dial complete: antibacterial facts & news. [Online.] http://www.dialsoap.com/index.cfm?page_id=104. Accessed 30 October 2005.

Thomas, M., W. Gillespie, J. Krauss, S. Harrison, R. Medeiros, M. Hawkins, R. Maclean, and K. F. Woeltje. 2005. Focus group data as a tool in assessing effectiveness of a hand hygiene campaign. *Am. J. Infect. Control.* **33**(6):368–373.

Tomes, N. 1998. *The Gospel of Germs: Men, Women, and the Microbe in American Life.* Harvard University Press, Cambridge, MA.

Tomes, N. 2000. The making of a germ panic, then and now. *Am. J. Public Health* 90(2):191–198.

Tomes, N. 2002. Epidemic entertainments: disease and popular culture in early-twentieth-century America. *Am. Lit. Hist.* 14(4):625–652.

U.S. Census Bureau. 2000. Census 2000 summary file 3 (SF 3)—sample data. [Online.] www.census.gov.

U.S. Department of Education. 2003. 2003 National Assessment of Adult Literacy. Institute of Education Sciences, National Center for Education Statistics. [Online.] http://nces.ed.gov/NAAL/index.asp?file=KeyFindings/Demographics/Overall.asp&PageId=16.

Wandersman, A. H., and W. K. Hallman. 1993. Are people acting irrationally? Understanding public concerns about environmental threats. *Am. Psychol.* 48(6):681–686.

Warner-Lambert. 2005. Listerine antiseptic mouthwash. [Online.] www.listerine.com. Accessed 31 October 2005.

Wedel, M., and W. A. Kamakura. 1999. *Market Segmentation: Conceptual and Methodological Foundations*, 2nd ed. Kluwer, Boston, MA.

Weinstein, N. D. 1984. Why it won't happen to me: perceptions of risk factors and susceptibility. *Health Psychol.* 3(5):431–457.

Weinstein, N. 1987. Unrealistic optimism about susceptibility to health problems; conclusions from a community-wide sample. *J. Behav. Med.* 10:481–499.

Weinstein, N. D. 1988. The precaution adoption process. *Health Psychol.* 7:355–386.

Williams, P. R. D., and J. K. Hammitt. 2001. Perceived risks of conventional and organic produce: pesticides, pathogens, and natural toxins. *Risk Anal.* 21(2):319–330.

Williamson, D. M., R. B. Gravani, and H. T. Lawless. 1992. Correlating food safety knowledge with home food preparation practices. *Food Technol.* 46:94–100.

Witte, K. 1998. Fear as motivator, fear as inhibitor: using the EPPM to explain fear appeal successes and failures, p. 423–450. *In* P. A. Andersen and L. K. Guerrero (ed.), *The Handbook of Communication and Emotion.* Academic Press, New York, NY.

Witte, K., and M. Allen. 2000. A meta-analysis of fear appeals: implications for effective public health campaigns. *Health Educ. Behav.* 27(5):591–615.

Woodburn, M. J., and C. A. Raab. 1997. Household food preparers' food-safety knowledge and practices following widely publicized outbreaks of foodborne illness. *J. Food Prot.* 60(9):1105–1109.

Worsfold, D. 1995. Recipe for food safety. *Nutr. Food Sci.* 95(6):22–25.

Worsfold, D., and C. J. Griffith. 1997a. Food safety behaviour in the home. *Br. Food J.* 99:97–104.

Worsfold, D., and C. J. Griffith. 1997b. Keeping it clean—a study of the domestic kitchen. *Food Sci. Technol. Today* 11:28–35.

Wynne, B. 1992. Public understanding of science research: new horizons or hall of mirrors? *Public Understand. Sci.* 1(1):37–43.

Wynne, B., and A. Irwin. 1996. *Misunderstanding Science? The Public Reconstruction of Science and Technology.* Cambridge University Press, Cambridge, United Kingdom.

Yang, S., F. J. Angulo, and S. F. Altekruse. 2000. Evaluation of safe food-handling instructions on raw meat and poultry products. *J. Food Prot.* **63**(10):1321–1325.

Young, J. H. 1992. *The Medical Messiahs: a Social History of Health Quackery in Twentieth-Century America.* Princeton University Press, Princeton, NJ.

Zhang, P., K. Penner, and J. Johnston. 1999. Prevalence of selected unsafe food-consumption practices and their associated factors in Kansas. *J. Food Saf.* **19**(4):289–297.

Index

A
Advertisements, appealing to emotions, 247–248
  beliefs about microbes and, 224
  concerning germs, 220–224
Advertising claims, for cellophane, 222
  for Clorox disinfecting wipes, 223
  for Crest Pro-Health Oral Rinse, 222
  for Dial soap, 222–223
  for Drano, 221
  for Electric Rat & Roach Paste, 221–222
  for Kleenex Anti-Viral tissues, 223
  for Lifebuoy soap, 221
  for Listerine antiseptic, 220–221, 222
  for Lysol wipes, 223
  for Mercurochrome, 221
  for Purell, 224
  for Raid with GermFighter Ant & Roach Killer, 223–224
  for "sneeze-proof" tissues, 221
Algae, 55–56
Antibiotic resistance, through food chain, case study of, 18–23, 24–26
Appropriate level of protection (ALOP), setting of, 35, 37–38, 43
Archaea, 60
Aspergillus spp., behavior of, 83
Audience, communicator understanding of, 211–213

B
Bacillus, 64
Bacillus cereus, 64, 88, 93, 95
  in vegetable puree, 119–124

Bacteremia, as cause of neonatal death, 180–181
Bacteria, 60–64
  modes of causing food-borne illness, 60–63
  spoilage, in foods, 78
Bacterial pathogens, food-borne, 61–62, 63–64
Behaviors, reflexive, practice of, 248
"Best before" date, 89
Bovine spongiform encephalopathy (BSE), 54
Broccoli puree production process, 120, 121

C
Campylobacter, 133, 152
  in chicken meat, 124–127
  in food, 230
Campylobacter spp., 95, 114
Case study: antibiotic resistance through food chain, 18–23, 24–26
Center for Food Safety and Applied Nutrition, 146
Ciguatera toxin, 55
Clostridium, 64
Clostridium perfringens, 64
Codex Alimentarius (Codex), 1
Codex Alimentarius Commission (CAC), 95, 143
  guidelines for microbial risk assessment, 65
  principles for microbial risk management, 144
  use of risk analysis by, 143

263

Codex Committee on Food Hygiene
    (CCFH), 143, 178
  assistance for, 143
Communication. *See also* Risk
    communication
  finding starting points for, 213–216
  successful, keys to, 210
Communicator(s), choice for risk communication, 209–216
  good, characteristics of, 209, 210
  understanding of audience, 211–213
Computer software systems, in predictive microbiology, 82
Consumer preferences, food consumption practices and, 245
Contamination, extrinsic, 187
Creutzfeld-Jakob disease (CJD), 54
Creutzfeld-Jakob variant (vCJD), 54
Cross-contamination, in communal kitchens, 232
Culture, and food risks, 226–227

D

Dating system, universal, on food packages, 243
Decision making, approaches to, 147–148
  regulatory, risk assessment in, 148–155
  technocratic, decisionist, and transparent models, 147–148
Disaggregation, 95
Disgust, as motivation for rejecting foods, 228
  indicators of, judgments about food safety and, 228–229

E

Emotions, appealing to, 248
Emulsions, water-in-oil, microbial growth in, 85
Encephalopathies, transmissable spongiform (TSE), 54
Enterobacter sakazakii, 142
  description of, 177
  dose required to produce illness, 182, 183–184
  ecology of, 178
  growth at various temperatures, 189, 190
  in infant formula, 177–204
    preparation and handling scenarios, 196–201
    prevention during manufacture, 193–201
  quantitation risk assessment for, 178
  reduction of risk of, setting of microbiological criteria and, 187, 193, 194–196
  risk characterization for, 191–192
  risk communication and risk management and, 201–203
  invasive infections from, as rare, 179–180
  exposure assessment of, 184–190
  hazard characterization in, 181–184
  hazard identification of, 179–181
  incidence of, 177
  outbreaks since 2001, 177–178
  ranges of concentrations of, 187
  risk analysis of, 178–179
  susceptibilities of infants to, 183–184
Environment, fluctuations in, predictive microbiology and, 86–87
Environmental factors, favoring growth of pathogens, 89
  microorganisms in foods and, 73
Environmental influences, on pathogen virulence, 90–93
Escherichia coli, 64, 92, 93
  in food, 230
European Food Safety Authority (EFSA), 144

F

FDA *Vibrio parahaemolyticus* risk assessments, 158–171
"Fear appeals," to influence behavior change, 246–247
Federal Register, 164
Food and Agriculture Organization/World Health Organization (FAO/WHO), on data needs and sources, 77–93
Food and Drug Administration (FDA), 30
  surveys of perceptions of food-borne illness, 240
Food-borne bacterial pathogens, 61–62, 63–64
Food-borne illnesses, communication concerning, as general, 241–242
  epidemiology of, changes in, 230–231
  incidence of, underestimation of, 239
  optimistic bias effects and, 239–242
  perceptions of, 240
    Food and Drug Administration surveys of, 240
  underreporting of, 240
Food-borne microorganisms. *See* Microorganisms in foods

Food categories, based on "cluster analysis," 168
　classification as "high" or "low" risk, 170
Food chain, antibiotic resistance through, case study of, 18–23, 24–26
Food consumption practices, and consumer preferences, 245
Food handling, unsafe practices in, high risk populations for, 235
　socioeconomic status and, 235–236
Food handling practices, and connection with risk, 239
　home-based, 231
　in home, food poisoning and, 241
　knowledge of, and implementation of, 237
　factors cited as barriers to practice, 238
Food hazards, health risks associated with, 51
Food hygiene training, ongoing reinforcement of, 242
Food industry, and environmental conditions in foods and during food processing, 52
Food labels, information on, 242–243
Food pathway, as chain of modules in modular process risk model (MPRM), 105–119
Food poisoning, recognition of symptoms of, 239
Food preparation, in home, investigations of, 231
　unrealistic optimism and illusions concerning, 240
Food preparation hygiene, failure to practice, 236–242
Food risks, affective responses to, 227–230
　culture and, 226–227
Food safety, risk assessment, microbial ecology in, 51–98
　risk assessment to establish, 39–42
　traditional process criteria for, 29–30
Food Safety and Inspection Service (FSIS), 30
Food safety issues, connection of actions to consequences in, 245–246
　increasing complexity of, 139
　messages concerning, appropriate channels for, 242–245
　television reporting of, 242
Food safety objective (FSO), 31–32, 38
　setting of, 43, 44–45, 47
Food Safety Objective (FSO)/Performance Objective (PO), 141
　paradigm, 34–39
　validation to meet, 47

Food safety problem, identification and prioritization of, 148–151
Foods, *Listeria* in, 225
　microbial ecology of. *See* Microbial ecology of foods
　microbial risks of, communicating about, 205–262
　scientific understanding for, 205
　microbial safety of, risk assessment principles in paradigm for controlling, 29–50
　microbiological hazards in, public perceptions about, 230–231
　microorganisms in. *See* Microorganisms in foods
　Salmonella in, 225
　spurned as disgusting, 228
　structure of, predictive microbiology and, 85–86
　undercooked, consumption of, 236
Fungi, 56

G

Gambierdiscus toxicus, 55–56
Gene, "expression" of, 92
"Germ theory," 219–220
Germs, advertising concerning, 220–224
　legacy of, 220–222
　and illness, beliefs about, 219
　educating consumers about, 221
　killing of, beliefs about, 219
　locations, beliefs about, 218
　messages about, in advertising, 224
　rare, concern about exposure to, 224

H

Hand sanitizers, waterless, 224
Hand washing, among health professionals, 235
　campaigns to encourage, 233
　food preparation and, 231
　reasons given for not, 233–234
　surveys in United States, 234–235
Hazard Analysis, Critical Control Points (HACCP), 30–31, 35
　adoption of, 140
　business/agencies using, 140
　data on explicit factors and, 78
　establishment of performance standards by, 140
　risk analysis by, 140
　uses of, 140

Hazard Analysis, Critical Control Points
(HACCP) *(continued)*
  verification of process operation according
    to, 47
Hazards, definition of, 139
Helminths, 57–58, 59
HHS/FDA and USDA/FSIS *Listeria monocytogenes* risk assessments, 158–171
HIV, maternal, infant feeding in, 182
Human health and safety assessments, 7
"Hurdle technology," 70

I

Import risk assessments, 7
Infant formula, powdered, cases of infection
  associated with, number of, 192
  contamination of, modes of, 187
  Enterobacter sakazakii in, case study
    using, 177–204
  handling, storing and usage methods
    for, 186, 187, 188
  microbial quality distribution of,
    191–192
  microbiological criteria for, 187
  prevention of contamination during
    manufacture, 193–201
  production of, 185, 186–187
  reconstitution of, temperature of water
    for, 198–200
  refrigeration of, 200–201
Infections, invasive, due to *Enterobacter
  sakazakii. See Enterobacter sakazakii,
  invasive infections from*
International Commission on Microbiological Specifications for Foods
  (ICMSF), 81
Interstate Shellfish Sanitation Conference
  (ISSC), 169

J
Jameson effect, 89–90

K
Knowledge deficit model, 215
Kuru, 54

L
Lactic acid, and pH, growth of microbes
  and, 74
Lactic acid bacteria, Jameson effect and, 90

Lag time(s), in small cell populations, 80
  information on, 79–81
  relative, frequency distributions of, 80–81
  variability of, 79
Lettuce, fresh-cut, food safety prarameters
  for, 37
Level of protection (LOP), 35
  appropriate (ALOP), 35, 37–38
  setting of, 43
Listeria, 225
  in food, 230
Listeria Action Plan, 170
Listeria monocytogenes, 64, 74–75, 90, 93,
  152
  food storage time and, 158
  HHS/FDA and USDA/FSIS risk assessments of, 158–171
  in foods at retail level, 164
  in ready-to-eat foods risk assessment, 157
  Jameson effect and, 90, 91
Listeriosis, 142
Literature review, and qualitative risk assessment, compared, 8

M
Mad cow disease, 54
Meningitis, as cause of neonatal death,
  180–181
Mibrobes, ecology of, beliefs about, 217–218
  knowledge of, sources of, 219–224
Microalgae, 55
Microbes, beliefs about, advertising and, 224
  decline in population and, patterns of, 66
  differences in lethality and, 65
  growth and death of, boundary between,
    70–71
  growth of, lactic acid and pH and, 74
  growth rate of, effects of temperature on,
    71, 72–73
  effects of water activity on, 71, 72–73
  population growth, patterns of, 69–70
  population inactivation kinetics, patterns
    of, 66, 67–68
Microbial contamination, in home kitchens,
  231
  knowledge of, 238
Microbial ecology of foods, 64–77
  changing patterns of microbial population
    and, 65–71
  deterministic aspects of, 53
  extrinsic factors in, 75–76
  factors affecting, categories of, 74

implicit factors in, 76, 81
in risk assessment of food safety, 51–98
interactions and correlations and, 88–93
intrinsic factors in, 74–75
processing factors in, 76–77
"rules governing," 81
safety risk assessments and, 93–96
Microbial food safety, regulatory decision making, using risk analysis for, 137–175
Microbial food safety issues, risk analysis for, reasons for, 139–143
Microbial food safety management systems, steps to improve, 139–140
Microbial growth, modeling of, 53
rate of population increase and, 65
Microbial interactions, predictive microbiology and, 86
Microbial loads, initial, estimation of, 78–79
Microbial pathogens, overview of, 54–64
virulence of, 92
Microbial risk assessment, Codex (Alimentarius Commission) guidelines for, 65
Microbial risk assessment model, 34–39
Microbiological criterion, 39
Microbiological hazards, in foods, public perceptions about, 230–231
Microbiological modeling, and risk assessment, 32–34
Microbiology, knowledge of peole about, 216–217
predictive. See Predictive microbiology
Microorganisms, effect of, and populations of microbes, growth, 52
in foods, responses of, 53
main groups of, 54–64
opportunistic, 33
pathogenicity of, determination of, 33
self-amplification of, 53
toxin production by, 33
virulence factors, and dose-response relationship, 34
and pre-consumption environment, 33–34
Microorganisms in foods, "biotic" environment and, 73
ecology of, predictive microbiology and, 81
environmental factors and, 73
lag time of, quantification of, 78–79
moving long chain, processes affecting number of, 73–74
1939 Milk Ordinance and Code, 29

Modular process risk model (MPRM), 94
and other risk assessment approaches, 127–129
basic processes in, 108–118
"black boxes" as steps in food chain, 116–118
building structure of, 104
collect necessary data and expert opinion, 104
cross-contamination in food safety and, 113–116, 117
define model to use for each module in, 104
description of food pathway in, 102–104
effects of growth, mixing, and cross-contamination in, 130, 132–133
experience with, 119–127
food-handling processes as removal in, 112–113
food pathway as chain of modules and, 105–119
growth of microorganisms and, 109
implement available data into model, 105
input and output of food chain model and, 118–119
methodology of, 102–105
microbial inactivation in, 109–110
mixing in, 111–112
model structure of food pathway processes, 120, 122
partitioning in, 110–111
perform exposure assessment in, 105
statement of purpose in, 102
steps to conducting QMRA with, 102
structured approach to food chain exposure assessment, 99–135
use for simpler food chain QMRA, 129–134
variability and uncertainty and, 106–108
Molds, 56–57
Mycotoxins, 56–57

N

National Research Council (NRC), and goals for risk communication, 206–207
"Negligible," to describe risk, 7

O

Organisms, specific spoilage organisms (SSOs), 78
Oysters, in risk assessments, 162, 163

P

Pathogens, environmental factors favoring growth of, 89
  virulence of, environmental influences on, 90–93
Pathogens in foods. *See* Microorganisms in foods
Performance criterion, 38–39
Performance objective (PO), 38
  setting of, 45
pH, and lactic acid, growth of microbes and, 74
Predictive microbiology, 81–83
  application of models to risk assessment, 83–88
  environmental fluctuations and, 86–87
  food structure and, 85–86
  microbial interactions and, 86
  model complexity and, 87–88
  strain variability and, 84–85
Prions, 54
Prokaryotes, 60
Proteins, required by cell, 92
Protozoans, 57–58, 59
Public perceptions, on microbiological hazards in foods, 230–231

Q

Qualitative risk assessment, 1–28
  and literature review, compared, 8
  and quantitative risk assessment, complementary, 4–5
  approaches to, 8–18
  case study from, 18–23, 24–26
  computation of final risk estimate in, 16
  conclusions of, 23, 24–26
  data needs for, 20, 21
  data presentation in, order of, 13–14
  data requirements for, 9–10
  defined risk question in, 8–9
  definition of, 1
  expert opinion and, 12
  identification of risk elements in, 14–16
  in food safety, 7–8
  main principles of, 2
  minimum requirements of, 3
  multiple data sets for, approaches to, 10–11
  numerical inputs into, 10
  omitting of irrelevant data from, 13
  reaching risk conclusions in, transparency in, 14–16
  reasons for specifying, 2
  risk assessment process in, 20–22
  risk pathway for, 19
  risk question for, 19
  steps in, 16
  tabular format for presenting data in, 14–16
  textual conclusions in, subjective nature of, 5–7
  uncertainty and variability and, 11–12
  use of, 2–4
Qualitative risk characteristics, 224–225
Quantitative microbial risk assessment (QMRA), 99
  conduction with modular process risk model, 102
  of food chain, 100
Quantitative risk assessment, as complementary, 4–5
  computation of final risk estimate in, 16
  knowledge necessary for, 9–10

R

Recipes, ingredients in, checking food safety hazards, 244–245
Refrigeration, need for, food labels and, 242–243
Refrigerator, temperature inside, 232
Regulon, 92
Review process, in risk assessment, 152–154
Risk, acceptable, 225–226
  as function of probability, 139
  comparative judgments of, 225–226
  dread, 225
  levels of, assessment of, 225
  words to describe, 6
  perceived, psychometric bases of, 224–225
  perceptions of, 224–226
  qualitative characteristics of, 224–225
  unacceptable, 225–226
  unknown, 225
Risk analysis, 39–40
  activities of, and responsible parties, 146
  benefits of, 141
  conditions for smooth operation of, 145–146
  expertise required for, 142
  financial resources for, 142–143
  for microbial food safety, using regulatory decision making, 137–175
  for microbial food safety issues, reasons for, 139–143
  framework for implementation of, 143–145

limitations of, 141–143
process of, 138
requirements of, 145
U.S. Federal Government in, 144–145
Risk assessment(s), 40–43
abstract scenarios for, 157 158
and risk characterization, 4
approaches for, 160–161
"cluster analysis" technique in, 168
data for, 165–166
data needs and sources for, 77–93
deciding whether to conduct, 151–155
decisions for, examples of, 155
definition of, 138
FDA *Vibrio parahaemolyticus*, 158–171
food safety, microbial ecology in, 51–98
foods of concern in, 162
future of, 171–172
  faster and better assessment in, 171–172
  improved communication in, 172
  improved transparency in, 172
heuristic scenarios for, 157
HHS/FDA and USDA/FSIS *Listeria monocytogenes*, 158–171
in regulatory decision making, 148–155
key results of, 166–169
microbiological modeling and, 32–34
organisms evaluated in, 161–162
plan for, 151
populations of concern in, 162–163
pragmatic scenarios for, 156
procedures for, 164–165
putting to work, 155–156
qualitative. *See* Qualitative risk assessment
questions to be answered by, 152, 153
refrigerator temperature scenario for, 157–158
review process in, 152–154
risk management options in, 169–171
stakeholder assessments and, 163–164
stakeholder involvement in, 154
storage times scenario for, 158
to analyze and improve food process, 44–47
to establish food safety, 39–42
trigger for, 159
two-dimensional, outcomes of, cumulative plots of, 45, 46
output of, 43
"weighted" data in, 165
Risk assessor, 6

Risk characterization, combining steps to form conclusions, 16–18
for *Enterobacter sakazakii* in infant formula, 191–192
"matrix" approach to, 18
risk assessment and, 4
Risk communication, and *Enterobacter sakazakii* in infant formula, 201–203
choosing communicator for, 209–216
common goals for, 206–209
definition of, 138
failure to clarify goals in, risks of, 207–208
"two-way" communications for, 208
understanding of science and, 214
Risk communicators, use "fear appeal" messages by, 247
Risk management, and *Enterobacter sakazakii* in infant formula, 201–203
definition of, 138
options in, identification and selection of, 154–155
quantitative microbial, activities associated with, 148, 150
seven components of, 148, 149
Risk manager, 6
CAC definition of, 143

S

Salmonella, 92, 133, 225
concerns about health and death associated with, 240–241
in food, 230
Salmonella enterica serovar Typhimurium, 93
Salmonella enteritidis, 152
Scrapie, 54
Seals, on food packages, explanation of, 243
Shelf life and storage conditions, 88–89
Shellfish, in risk assessments, 162, 163
Sigma (σ) factors, 92
Specific spoilage organisms (SSOs), 78
Staphylococcus aureus, 93
Staphylococcus xylosus, 71

T

Television, as potential source of food safety information, 244
Transmissible spongiform encephalopathies (TSE), 54

U

U.S. Federal Government, in risk analysis, 144–145

U.S. OMB Peer Review Bulletin, 152
USDA/FSIS and DHHS/FDA, regulatory responsibilities of, 144–145
"Use by" date, 89

V

Validation, to meet Food Safety Objective (FSO)/Performance Objective (PO), 47
Verification, of process operation according to Hazard Analysis, Critical Control Points (HACCP), 47
Vibrio, 64
Vibrio parahaemolyticus, 152
  FDA risk assessments of, 158–171

Vibrio spp., 93
Viruses, 58–60

W

Waterless hand sanitizers, 224
World Organisation for Animal Health (OIE), 1
World Trade Organization, Sanitary and Phytosanitary Agreement of, 35–36

Y

Yeasts, 56
Yersinia enterocolitica, 93